Information Theory
as Applied to
Chemical Analysis

CHEMICAL ANALYSIS

A SERIES OF MONOGRAPHS ON ANALYTICAL CHEMISTRY AND ITS APPLICATIONS

VOLUME 53

A WILEY-INTERSCIENCE PUBLICATION

JOHN WILEY & SONS

New York / Chichester / Brisbane / Toronto

Information Theory as Applied to Chemical Analysis

KAREL ECKSCHLAGER

*Institute of Inorganic Chemistry
of Czechoslovak
Academy of Sciences, Prague*

VLADIMÍR ŠTĚPÁNEK

*Environmental Research Centre,
Prague*

A WILEY-INTERSCIENCE PUBLICATION

JOHN WILEY & SONS
New York / Chichester / Brisbane / Toronto

Library of Congress Cataloging in Publication Data

Eckschlager, Karel.
  Information theory as applied to chemical analysis.

  (Chemical analysis; v. 53)
  "A Wiley-Interscience publication."
  Includes index.
  1. Chemistry, Analytic—Mathematics. 2. Information
theory. I. Štěpánek, Vladimir, joint author. II. Ti-
tle. III. Series.
QD75.3.E36   543'.001'5192   79-1405
ISBN 0-471-04945-X

Printed in the United States of America

10 9 8 7 6 5 4 3 2 1

PREFACE

In teaching analytical chemistry, in researching new methodologies of analytical chemistry, and in doing practical analytical work, there is need for individual analytical methods that will rapidly and economically provide necessary information about the composition of an analyzed sample. The concept of "information" allows us to derive quantities that facilitate evaluation (and classification) of analytical methods, make them more objective, and serve as objective functions in optimizing analytical procedures.

This book discusses the fundamentals of the current understanding of information concepts in analytical chemistry and aims at presenting most of their applications described so far. This objective appears in the arrangement of the text. Our intention has been to make the monograph self-contained as far as the employed theory is concerned. The mathematical chapters require only knowledge of calculus and of combinatorial counting techniques. Instead of a formal presentation of the concepts we have tried to explain them without including most of the proofs. We have emphasized notation because much confusion has arisen from imprecision in its use.

The introductory chapter is followed by Chapter 2, in which the basic ideas of understanding analytical chemistry as a process of obtaining information are explained. Since messages bearing information are random, randomness must be the basis of every realistic analysis. In this light probability theory is the only tool that can be used to pursue it. Chapter 3, therefore, is devoted to the fundamental concepts of this discipline. Information theory, which has developed from probability theory, is the theoretical study of information measures and information transmissions. It is discussed in Chapter 4. The first goal of this theory is to propose measures of information content and information gain. Since

analytical methods are verified by statistical means using experimental results and various statistical techniques are adopted in the
text, we have included a chapter (Chapter 5) dealing with some
methods of statistical inference. The information access to the
evaluation of various procedures in analytical chemistry is examined in the last chapter, where a survey of recent results
appears. The results are discussed and possibilities of practical use
are shown. Some information measures, introduced in this chapter
and illustrated on examples, are specific for the evaluation in
chemical analysis.

Each chapter is divided into sections and sometimes subsec-
tions. Examples are intermingled with the text. References are
given at the end of each chapter in the sequence in which they
appear in the text. Some of the references are not cited in the text.

We wish to thank those who stimulated the writing of this book.
We feel indebted to Professor J. D. Winefordner of the University
of Florida, Gainesville, for his interest in this monograph for the
Chemical Analysis Series. We thank him and Professor J. P.
Elving, the other editor of the series, for their help and encourage-
ment during the preparation of the book. We appreciate suggestive
discussions of some theoretical problems with Dr. I. Vajda. We
also thank Mrs. J. Vyvialová for the remarkable speed and accuracy with which she typed the manuscript.

KAREL ECKSCHLAGER
VLADIMÍR ŠTĚPÁNEK

Prague
June 1979

CONTENTS

GLOSSARY

CONVENTIONS

We have used Greek letters (except in a few cases for historical reasons) to represent random variables and the corresponding Roman letters to represent their values. Thus, x are values of a random variable ξ. Similarly, a clear distinction is drawn between parameters and estimates of these parameters calculated from observations. Thus, θ denotes a general parameter of a probability distribution and $T = T(x_1, x_2, \ldots, x_n)$ is its estimate calculated from a sample of size n.

Observations of one random variable are indexed by one subscript. If we deal with two or more random variables, observations are indexed by two subscripts; that is, y_{ij} is the jth observation of the random variable η_i.

Several special conventions are used, primarily in Chapters 3 and 4.

Symbols

$A = \dfrac{\sigma}{\sigma_0}$	Coefficient of precision making
a	Estimate of the regression coefficient α
a_0, a_k	Coefficients of the Fourier series
α	Probability of a type I error (the significance level); regression coefficient
A_P	Plane-discrimination capability
A_V	Space-discrimination capability
$B = \dfrac{\mu - \mu_0}{\sigma_0}$	Coefficient of modification of the results
b	Half-width of a peak; estimate of the regression coefficient β
b_c	Slope of a calibration curve

b_0, b_k	Coefficients of the Fourier series
β	Probability of a type II error; a regression coefficient
$C(p,p_0)$	Specific information price
c_i	Concentration (content) of the ith component
d_i	Estimate of the mean error of the determination of the ith component
δ	Mean error
$\delta^{(0)}$	Mean error related to blank correction
$E[\ \ ]$	Expectation of the quantity in brackets
e_i, ε_i	Effectivity coefficients
η_i	Signal intensity for the ith component
F	Random variable
F_α	100α-percentage value of the F-distribution
$f_i(X_i)$	Function of the true value of the ith component (known or found by calibration)
$g_i(y_i)$	Function of the signal intensity of the ith component (inverse of the function f_i)
$H(x_1,x_2,\ldots,x_n)$	Joint distribution function of n random variables
H, H_0	Statistical hypotheses
i	Subscript i, index of the component or of a random variable
$I(p,p_0), I(q\|p)$	Information content; the gain of information
$J(p,p_0)$	Information flow
j	Subscript j, index of a parallel determination
K	Tolerance coefficient
k	Number of (determined) components
$L(p,p_0)$	Information performance
$\ln, \log$	Logarithms
λ	Parameter of the Poisson distribution
M	Total number of components in an analyzed sample
m	Number of components detected by a given qualitative test; weighed amount of a sample; number of signals

μ_i	Expected value of the random variable ξ_i, $\mu_i = E[\xi_i]$
N	Number of elements of a finite set
N_1	Number of possible discriminations of the substance
N_2	Number of discriminated concentrations
$n(A)$	Number of sample points of the event A
n_p	Number of parallel determinations
n_s	Number of determinations from which an estimate of the standard deviation is calculated
$\binom{n}{k}$	Number of combinations of n elements taken k at a time
ν	Number of degrees of freedom
$\Omega_{i-1,i}$	Overlap between the $(i-1)$th and the ith peaks
$P(A)$	Probability of the event A
$P(A\|B)$	Conditional probability of event A given event B
p_i	Probability of a discrete random variable, $p_i = P\{\xi = x_i\}$
$p_0(x)$	Probability density of an a priori distribution
$p(x)$	Probability density of an a posteriori distribution
q_i	Probability of a discrete random variable, $q_i = P\{\eta = y_i\}$
q_0	Ratio of the peak symmetry
q	Ratio of the peak shape (for symmetrical peaks only)
$\mathscr{R}_1$	Set of real numbers
ρ	Information redundance
S	Set of all possible outcomes of a random experiment
S_i	Sensitivity of determination of the ith component
s	Estimate of the standard deviation
s_y	Estimate of the standard deviation of the signal intensity

σ	Standard deviation of a random variable
σ^2	Variance of a random variable
σ_i^2	Variance of the proper determination of the ith component
σ_0^2	Variance of the blank correction; variance of an a priori probability distribution
T	Period in a Fourier series
t	Student's random variable
$t_\alpha(n)$	100α-percentage value of the Student distribution with n degrees of freedom
t_ν	Percentage value of the Student distribution for $\alpha = 0.038794$ and $\nu = n_s - 1$ degrees of freedom
t_A	Time of the duration of an analysis
τ_A	Cost of an analysis
θ	General parameter of a probability distribution
$V[\xi]$	Variance of the random variable ξ
X_i	True content of the ith component in an analyzed sample
$x_{i,j}$	Value of the result of analysis of the ith component and of the jth determination
$x_i^{(c)}$	Amount of the component to be determined obtained from the calibration curve
$\bar{x}_i$	Mean of parallel determinations of the ith component
x_U	Upper limit of a tolerance interval
x_L	Lower limit of a tolerance interval
ξ_i	Results of analysis of the ith component
y_i	Value of the signal intensity for the ith component
$\bar{y}_i$	Mean signal intensity of the ith component
y_{min}	Least intensity of the signal distinguishable from zero
y_{max}	Maximum intensity of the signal
z_i	Signal position for the ith component
z_α	100α-percentage value of the standardized normal distribution

$$\sum_{i=1}^{n} x_i = x_1 + x_2 + \cdots + x_n$$

$$\prod_{i=1}^{n} x_i = x_1 x_2 \cdots x_n$$

$\in$ Element of

$\cap$ Intersection of sets

$\cup$ Union of sets

Information Theory
as Applied to
Chemical Analysis

INTRODUCTION

Analytical chemistry, the scientific discipline representing theoretical grounds for chemical and physicochemical analyses of the composition of matter, has in the last few decades been characterized by increasing interest in problems of more general validity. In the first period of development of this scientific discipline, "analytical chemistry" usually was understood to be a collection of procedures to be carried out when performing an analysis. In 1977, W. Fresenius presented this definition of analytical chemistry (paraphrased from [1]): "Analytical chemistry is the science of acquiring information on material systems and interpreting it with regard to its exploitation, employing the methods of natural science." In this definition obtaining information is said to be the objective of chemical and instrumental analyses. We can then define the analysis itself as a process of obtaining information about the chemical composition of matter. From it also follows the importance of the use of concepts and methods of information theory for analytical chemistry.

The development of the concept of analytical chemistry from the original idea of a collection of working recipes to the present "science of acquiring information" was of course quite slow.

An important development was H. Kaiser's use of the probabilistic point of view in 1936 in connection with the poorly reproducible emission spectrometry results available at that time [2]. The basis of the probabilistic point of view resides in the idea that the result of the determination of the unknown content of a component is a random variable which has a probability distribution. This idea enabled evaluation of analytical results and methods by the use of statistical methods. This became quite common in analytical practice during the 10 to 15 years following the first papers by Kaiser [2]. Statistical evaluation of analytical results and methods has spread primarily with the development of instrumental methods, especially trace analysis. There have also appeared probability definitions of purely analytical concepts: for example, Kaiser has presented a statistical definition of the detection and determination limits and J. D. Winefordner and later L. A. Currie have extended it. In the literature of trace analysis, concepts from communication theory are commonly used (e.g., signal, signal-to-noise ratio, detection of signals, etc.). In connection with introducing automation and machine processing of analytical data, an "automation in analysis" work team was established in Lindau at the beginning of the 1970s, and its members have contributed many new approaches and given a number of useful definitions. This group, in their wide and generally directed activities, have also been concerned with questions of the use of information theory in analytical chemistry. Here again, the probabilistic approach is evident, since the concept of information is narrowly linked with the probability distributions of appropriate random variables. Although Kaiser was not the first to be concerned with the possibilities of using information theory in analytical chemistry, his lecture delivered at the University of Georgia in 1969 and published one year later [3] had fundamental importance for the extension of the theoretical point of view of information theory in the literature of analytical chemistry. A certain gain in the understanding and practical use of the information content of analytical results has come with the introduction of the divergence measure. This has substantially more general validity than have measures formerly transferred from communication theory [4].

In the last one or two decades new theoretical and often mathematically expressed approaches to analytical problems have appeared in the literature of analytical chemistry. These new approaches are often accompanied by the introduction of new technology for practical analyses. This is primarily done by transferring basic theoretical notions from other fields of science and

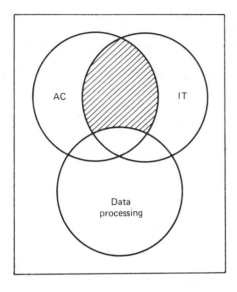

Figure 1.1 Venn diagram. AC, analytical chemistry; IT, information theory.

adapting them for the needs of analytical chemistry. In this way analytical chemistry has become a multidisciplinary field: information theory emerging from probability theory today forms a part. Information theory had its origins in the 1920s, when R. A. Fisher introduced some basic definitions; it spread extensively in the United States after 1945, through the work of N. Wiener, C. E. Shannon, S. Kullback, and others.

In this book, we will be concerned with the use of information theory to describe, evaluate, and optimize processes of obtaining information in analytical chemistry. We will not pay attention to automation and data processing, although information theory can be utilized in these fields as well. The sphere that is the object of our interest is shown by the Venn diagram of Figure 1.1.

All information theory is based on the notion that information and uncertainty are synonymous. Since probability theory is the mathematical study of uncertainty, it is fundamental to information theory. Without it, the central notions of information theory could not be pursued.

References

1. W. Fresenius, *Reviews on Anal. Chem.* p. 11, Akad. Kiadó, Budapest, 1977.

2. H. Kaiser, *Z. Tec. Phys.* **17**, 219, 227 (1936).

3. H. Kaiser, *Anal. Chem.* **42**, No. 2, 24A; No. 4, 26A (1970).

4. K. Eckschlager, *Z. Anal. Chem.* **277**, 1 (1975).

ANALYSIS AS A PROCESS OF OBTAINING INFORMATION

2.1 THE ANALYTICAL PROCESS

Chemical analysis is a process used to obtain information about the chemical composition of an analyzed sample or material from which the sample was taken. Although the analysis may be performed for various purposes (e.g., in quality control of raw materials and products, medical diagnosis, geological prospecting, investigations of environmental pollution), the information obtained by the analysis is always in answer to the questions "What?" or "How much?". The first question is answered through *qualitative analysis*, the second through *quantitative analysis*. By the use of modern instrumental methods we can obtain both answers simultaneously. Sometimes we have to combine a few analytical methods to obtain the required information. The results of qualitative analysis are by nature alternative information: they indicate if a certain expected component is or is not present in the sample or, better, if it is or is not present in an amount higher than the detection limit of the procedure used. The results of quantitative analysis are characterized by numerical information; they may be expressed as values of concentration or content of a certain component in the analyzed sample.

The process of obtaining information about the chemical composition of a substance has two steps. In the first, information is created and coded into an analytical signal; in the second, it is decoded and transferred into a result. This process is shown in Figure 2.1. Both steps are necessary to obtain finite information,

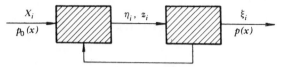

Figure 2.1 Scheme of quantitative analysis I.

and the steps run independently, and a negative feedback can be created. If the process of obtaining information about the chemical composition of a substance is to run effectively, the levels of the functions of both systems have to be at least approximately balanced.

The first step, the *experimental creation of the information,* occurs in a real system (e.g., in an analytical device). A sample enters this system (together with the inevitable reagents and energy), which has a fixed but unknown content of component X_i to be determined (where $i = 1, 2, \ldots, k$; k is the number of all components to be proved or determined). The output from this system is an analytical signal that is always realized through a certain change in physical state (e.g., change of electrical current or voltage).

The analytical signal always has a certain intensity η_i, which takes on a value y_i and depends on the content of the ith component, X_i, in the sample; thus it contains the answer to the question "How much?". The signals may have different positions for different components $i = 1, 2, \ldots, k$; and this position contains the answer to the question "What?". Since analytical signals usually appear with noise in the background, we consider only signals of intensity $y_i \geqslant y_{\min}$, where $y_{\min}$ is the smallest value of the signal intensity which is distinguishable from zero. The nature of an analytical signal intensity η_i is always that of a random variable,* and therefore we usually consider the dependence of its mean value $E[\eta_i]$ on X_i, that is, the dependence

$$E[\eta_i] = f_i(X_i) \tag{2.1}$$

*See Chapter 3 for this and other concepts of probability theory.

Here the function f_i is either known (e.g., from stoichiometry) or we determine it empirically (i.e., by calibration). The individual mutually distinguishable positions of the signal z_i correspond to different components of the analyzed sample ($i = 1, 2, \ldots, k$). The value z_i for a certain component will be taken as a fixed (deterministic) variable (e.g., the wavelength of a spectral line)—or at least the rank in which signals corresponding to individual components occur always remains unchanged for a given analytical procedure. In fact, for some methods of structural analysis the position of the signal z_i can be a random variable as well. Sometimes a sequence of signals is called a "message," obviously according to the pattern of models known from communication theory.

The second step in obtaining information about the chemical composition of a substance, the *decoding of an analytical signal*, occurs either in a conceptual (logical) system or in a real system (e.g., in a computer). The output from this system is a result ξ_i, which takes on values $x_{i,j}$ and includes both information about the qualitative composition (the decoding of the position z_i) and numerical information indicating the content found for the ith component determined (the decoding of the intensity y_i).

Since the result ξ_i arose by decoding the random variable η_i (i.e., the intensity of the signal), it is itself a random variable. The decoding of an analytical signal is a process which is important to the reliability and information content of the result. If the decoding is not carried out completely, loss of a part of the information created in the first system takes place; but even perfect decoding of the analytical signal cannot give rise to information not contained in the analytical signal being processed. In customary analytical practice, we often decode only information that is useful for the given purpose. Sometimes, one system can decode analytical signals entering it from several systems; for example, a computer working in a time-sharing regime can be connected "on-line" to several analytical devices at the same time.

Negative feedback can be created between both systems by which the second system controls the action of the first system so that its function is optimal for the given analytical sample. Such control of a system creating information (e.g., of an analytical device) can be performed experimentally by a person who adjusts the conditions of the determination according to preliminary results (e.g., the weight of a sample, the sensitivity of analytical apparatus) so that they are as optimal as possible. Control of the experimental system can also be automatic when, for example, a computer connected on-line to the analytical device under working conditions in real time controls its operation by means of feedback so that the device works fully automatically under optimal conditions.

2.2 QUALITATIVE AND QUANTITATIVE ASPECTS OF ANALYTICAL INFORMATION

Information is being defined as the decrease in uncertainty of our knowledge of an object, event, or action. All information, including analytical information, can be viewed from two viewpoints: qualitative (i.e., it has a certain relevance for recognizing the object or action), and quantitative (given by the magnitude of the decrease in the uncertainty of our knowledge of the chemical composition of the sample reached through analysis). We cannot determine the relevance of analytical information for a certain concrete purpose in numerical form expressed in certain units. However, we will show later that it is possible to choose an analytical method or adjust an analytical procedure so that when we obtain information about the presence or content of several variables at the same time, information highly relevant for the given task is obtained with the greatest possible decrease in uncertainty. The most relevant information is obtained simultaneously with the most comprehensive information.

For evaluation of the quantitative part of the analytical information we have to define uncertainty before performing the analysis and afterward find an appropriate measure that might characterize this part of the obtained information. We will follow the model of analysis as a process of obtaining information about the chemical composition of the sample as described in Section 2.1. In defining uncertainty before performing the analysis, we can imagine that we have a certain preliminary knowledge of the sample entering the first system (an analytical device), or that we set assumptions regarding the qualitative or quantitative composition of the sample. Without this preliminary knowledge or assumption it would not be possible to carry out the analysis, for we could not even choose a suitable analytical method. Of course, this preliminary a priori knowledge always has some uncertainty (sometimes quite large), which is described by a probability density or a density function $p_0(x)$. The result of the analysis, as a random variable, contains in it some uncertainty as well; this a posteriori uncertainty is expressed by a probability density or a density function $p(x)$ (see Figure 2.1). These distributions differ somewhat: the a priori distribution $p_0(x)$ characterizes our ignorance of the fixed but unknown value X_i, whereas the a posteriori distribution $p(x)$ expresses uncertainty caused by the result ξ_i being a random variable. We will see later that the quantitative aspect of the information contained in the result of the analysis is conditioned only by the a posteriori distribution $p(x)$ and by the preliminary a priori distribution $p_0(x)$, without respect to what measure we use to express the quantitative part of this information.

In connection with development of the theory and practice of rational storing and automatic processing of analytical information, some concepts and methods of semiotics, the scientific field dealing with studies of symbol systems, grow in their importance to analytical chemistry. Semiotics has three parts: (1) syntax, concerned with the internal structure of the symbol systems and with the relations of the symbols of a certain system among themselves; (2) semantics, concerned with the symbol system as a means of expressing the sense of a message and with the relations of the symbol to what it expresses; and (3) pragmatics, dealing with the relation of the symbol to the addressee of the information. These problems are dealt with in papers by H. Malissa [1].

2.3 PROPERTIES OF THE PROCESS OF PRODUCING INFORMATION ABOUT THE COMPOSITION OF THE SAMPLE

The system producing information about the chemical composition of the sample (e.g., an analytical device) works such that for a certain interaction of the analyzed sample with the reagent and/or energy, it produces a signal at its output whose shape and possibilities of being decoded are conditioned by the properties of the system. In analytical practice we judge the systems producing information according to a number of properties. Various properties of the systems producing analytical information have been discussed in detail in recent scientific literature; these studies sometimes give somewhat differing definitions and quantitative relations expressing these properties. Here we will discuss only those that are important for this text and will adopt certain descriptions for them even if our descriptions differ from those used in other publications. Among the properties that we will pay attention to are sensitivity, specificity and selectivity, linearity, blank correction, signal-to-noise ratio and detection limit linked with it, time dependence of the signals, time stability of the zero, and the stability of the noise of the zero.

Sensitivity characterizes a change in the response of the device output to a change in the concentration of the measured component at the input. If the dependence of the intensity of the signal η_i on the true content X_i of the component to be determined is given by the relation (2.1), that is, if $E[\eta_i] = f_i(X_i)$, the sensitivity is defined as

$$S_i = \frac{dE[\eta_i]}{dX_i} = \frac{df_i(X_i)}{dX_i} \tag{2.2}$$

The value of S_i increases with increasing rate of change of the average intensity of the signal with the change of the concentration or the content of the ith component to be determined. Some systems producing information about the chemical composition of the sample (i.e., some analytical procedures or devices) have variable sensitivities; one can sometimes change the sensitivity continuously in the range of three to four decades. The feedback between the systems in the process of obtaining information is usually based on regulation of the sensitivity of the system in which the information is produced.

The *specificity* is a property of the system producing the analytical information characterized by the dependence (2.1) holding only for one component [e.g., for the component $i=1$ if the analyzed sample contains k components $(k>1)$]. Therefore, $E[\eta_1]$ $=f_1(X_1)$, whereas such a function does not exist for $i=2,3,\ldots,k$ (i.e., the system responds only to the content of the first component). In practice we must often contend with "relative specificity," where $S_1 \gg \sum_{i=2}^{k} S_i$, that is, when the sensitivity of the device or the magnitude of the response of the device to the same concentrations of the components $i=2,3,\ldots,k$ is entirely negligible relative to the sensitivity of the response to the first component. The specificity is thus a property of the system producing analytical information that enables determination of one component of the sample in the presence of others.

Selectivity is a more advantageous property than specificity; it enables determination of several components of the analyzed sample simultaneously. Simultaneous determination of several components is, of course, feasible only if the output of the first system is formed by a sequence of mutually distinguishable signals. The number of signals is $m \geqslant k$, and there corresponds at least one analytical signal, whose average intensity $E[\eta_i]$ is a function of the content X_i of the component to each component $i=1,2,\ldots,k$. The difference between the positions of the signals must be so large that we can independently determine intensities of all the signals, and the sensitivities of the determinations of individual components must be at least approximately equal. The problems of selectivity and specificity of analytical methods have been dealt with by Kaiser [2].

In methods of local analysis where a plane or space coordinate of the surface of the sample comes as an additional variable, another very important property, the *geometrical* (plane or space) *discrimination capability* is considered. This is defined as

$$A_V = \frac{V}{\Delta V} \quad \text{or} \quad A_P = \frac{P}{\Delta P} \tag{2.3}$$

where P and V are the total area or volume of the analyzed sample which is treated by local analysis. It does not matter whether we work with a point analysis or with a scanning method; ΔP and ΔV are the smallest area or volume that we can analyze. The geometrical discrimination capability A thus always represents the number of possible space or plane discriminations that we can carry out. The numerical value A_V is usually related to $V = 1$ cm^2 and A_P to $P = 0.1$ cm^3.

Linearity is an advantageous but not essential property of the system which produces information about the composition of the matter. It occurs whenever the response of the output of this system $E[\eta_i]$ is a linear function of X_i. We define linearity within the bounds of the content of the component to be determined, $X_i \in \langle X_{i1}; X_{i2} \rangle$, as in the case when

$$E[\eta_i] = \alpha + \beta X_i \tag{2.4}$$

This is the case of linear regression (see Section 5.5 for its analysis). If the sensitivity given in (2.2) decreases in this interval as the true content X_i increases, this dependence is concave (curve 1 in Figure 2.2). Inversely, if the sensitivity increases with decreasing contents X_i, the dependence is convex (curve 2 in Figure 2.2). Frequently, the dependence of the response (i.e., of the average signal intensity $E[\eta_i]$ on X_i) is linear only in a domain of the values X_i and shows deviations from linearity outside it. The

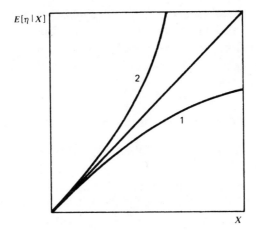

$E[\eta \mid X]$

X

Figure 2.2   Linearity and nonlinearity of the output: (1) concave curve; (2) convex curve.

regression coefficient α in (2.4) can be set zero by blank correction if it is due to contamination. Such a blank correction influences, of course, the precision of the result (see Section 6.4 and [6]). The problem of blank corrections in the case of a nonlinear dependence has been discussed by van der Linden [7].

Next properties will rather concern analytical instruments. Among them, first of all, the *signal-to-noise ratio* determines, along with the sensitivity (2.2), the detection limit in trace analyses [6, 8–10].

The *time independence* (the stability) is a very important property of a system in practice. It is defined as the case where the intensity of the signal η_i for a certain fixed concentration X_i does not depend on time. We are often contented with η_i being constant in the time interval $\langle t_0, t_D \rangle$, where t_D is the mean period for which the device daily operates. In practice, the *time stability of the zero* (i.e., the case when $y_0 = f_i(0) = $ constant in time, at least between t_0 and t_D) is also important. For trace analyses, where we obtain small values of the intensities of the signals, the *time stability of the noise* (i.e., the time independence of the parameter $V[\eta_i]$) and, especially, the *time stability of the noise of the zero* (i.e., of $V[\eta_i]$ for $X_i = 0$) are also important. It must be at least approximately constant and not too large. Experience shows that the noise of the signal intensity $V[\eta_i]$ depends to an extent on the sensitivity defined by (2.2).

2.4 PROPERTIES OF AN ANALYTICAL SIGNAL

An analytical signal is realized by a change of physical state, most frequently by a change in electrical current or voltage. The decoding method employed depends primarily on whether we are dealing with a one-dimensional signal, from which we determine only its intensity, or with a two-dimensional signal, for which we determine both its intensity and its position. For spectral methods H. Kaiser [5] has termed the one-dimensional methods monochromatic and the two-dimensional methods polychromatic.

A *one-dimensional signal* can be a digital value which we read from a digital display or an analog value (the position of the hand on the scale of the measuring device or of the pen on the recorder), or it can be formed by a two-dimensional graph from which we determine only one value (i.e., the signal intensity). So, for instance, from two-dimensional titration curves we are concerned only with the consumption of the volumetric standard solution. Similarly, we have to consider as one-dimensional that signal whose position (e.g., in the graphic record) contains no information concerning the analyzed sample; for example, the distance of rectangular signals obtained in routine use of the atomic absorption spectrometry is given by the rate of the exchange of the samples and is thus not "a position" of the signal in the sense of our interpretation in Section 2.1. The intensity of the signal is given in the graph, for example, by the length of the distance from a "fixed" point (often from the "zero" point) or from the baseline to a certain point on the curve. From a practical point of view it is important how sharply this point is indicated on the appropriate curve. In titration curves it is the slope of the curve at the inflection point that indicates the equivalence point; for peak-shaped signals, the sharpness and symmetry are important. However, in any case the intensity of a signal η_i is always a random variable.

A *two-dimensional signal* (i.e., a sequence of individual signals) is either formed by a two-dimensional graph, from which we determine positions and intensities of individual signals, or it is given by pairs of digital values, from which one always characterizes the position and the other the intensity of the signal. The analog two-dimensional signal supplied, for example, by a recorder frequently has the shape of a step curve or the shape of a sequence of peaks. Sometimes we can transfer one shape into the other (by "differentiating" the step curve we obtain the peaks and by "integrating" the peaks, we obtain the step curve). The step curve is known from classical polarography or gas or liquid chromatography as the output from a recorder connected to an integrator. A digital record is also possible (e.g., on a paper or magnetic tape). On a graph of the step curve the distance from the zero point of the curve (which, of course, need not coincide with the origin of the curve) to the inflection point on the curve expresses the position of the signal z_i, and the intensity y_i is given by the magnitude of the degree of the curve. For sequences of signals of peak shape as we obtain them in many analytical methods, the intensity of the signal y_i is given either by the height or the area of the peak. For symmetrical peaks we sometimes distinguish Lorentzian, Gaussian, and rectangular shapes ("profiles"). The Lorentzian shape is given by

$$y = \frac{y_{max}}{1 - (2z/b)^2} \qquad (2.5a)$$

and the Gaussian shape by

$$y = y_{max} \exp\left[-\left(\frac{2z^2}{b} \right) \ln 2 \right] \qquad (2.5b)$$

where z is the difference $|z_i - z_{max}|$ and z_{max} is that value of z_i for which $y = y_{max}$ (i.e., it takes on its relative maximum for the given peak). The parameter b is the half-width of the peak and represents its width at $y = \frac{1}{2} y_{max}$. In the Gaussian function $b = 2\sigma\sqrt{2\ln 2} \approx 2.35\sigma$, where σ is the value z of the inflection point of the curve. An extreme case is that of the rectangular peak, where

$$y = \begin{cases} y_{max} & \text{for } z \in \langle z_{max} - \frac{1}{2}b, z_{max} + \frac{1}{2}b \rangle \\ 0 & \text{for other values of } z \end{cases} \qquad (2.5c)$$

Sometimes we get peaks of profiles other than those corresponding to rectangular, Gaussian, or Lorentzian shapes; we can characterize their profiles by the ratio

$$q = \frac{c}{a} \qquad (2.6a)$$

(see Figure 2.3). The values q for rectangular, Gaussian, and Lorentzian peak shapes are tabulated in Table 2.1. Sometimes we

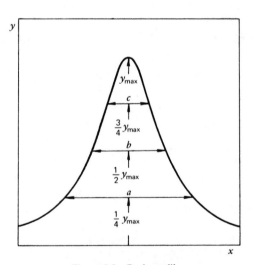

Figure 2.3 Peak profile.

TABLE 2.1 Values $q = c/a$ for Signals of Various Shapes

Shape of the Signal	q
Rectangular	1.000
Gaussian	2.196
Lorentzian	3.000

get asymmetrical peaks; we then characterize the symmetry (according to Doerffel [3]) by the ratio

$$q_0 = \frac{b_1}{b_2} \tag{2.6b}$$

where $b = b_1 + b_2$ (see Figure 2.4). The symmetry, shape, and half-width have a meaning for distinguishing two neighboring and partially overlapping signals. The possibility of distinguishing such signals is meaningful with regard to the selectivity of the determinations (Section 2.3) even when the selectivity is not given only by the discrimination capability of the device.

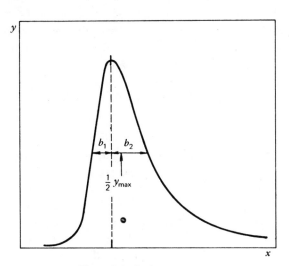

Figure 2.4 Peak asymmetry.

Cyclically recurring signals consisting of various peaks or having a shape arising from a combination (supposition) of several peaks can sometimes be described in terms of a Fourier series. Any continuous periodic function $f(x) = f(x + kT)$ with a continuous derivative, where T is a period, can be expanded in a convergent trigonometric series

$$y = \tfrac{1}{2}a_0 + \sum_{k=1}^{\infty} (a_k \cos kx + b_k \sin kx) \qquad (2.7)$$

with coefficients

$$a_k = \frac{2}{T} \int_0^{T_0} y \cos kx \, dx \qquad b_k = \frac{2}{T} \int_0^{T_0} y \sin kx \, dx$$

for $k = 0, 1, \ldots, \infty$. If the graph of the function is symmetrical, $b_k = 0$ and the series contains only members with the cosines. For details, see [4].

Distinguishing between one- and two-dimensional signals loses its importance somewhat if we decode them by a computer connected on-line to an analytical device. It also has importance in this case for programming the method of decoding the analytical signal. Besides this, as we will show, distinguishing between signals is important when judging the amount of information contained in the results of an analysis. Here, of course, we will be concerned rather with deciding whether we can determine only one component of the analyzed sample (a one-dimensional signal) or several components simultaneously (a two-dimensional signal).

2.5   THE DECODING OF INFORMATION ABOUT THE COMPOSITION OF SAMPLE

Information about the chemical composition of an analyzed sample stored in an analytical signal which leaves the first experimental system is decoded in a real system (a computer) or in a logical conceptual system. Through decoding we obtain the result of the analysis.

The result of a *qualitative analysis* contains, by its nature, alternative information (e.g., the ith component is or is not present in the sample or its content in the sample is greater or less than that corresponding to the detection limit of the method used). Formally, we can write the decoding or evaluation of such a qualitative finding in the form of an operator equality:

$$i = A\{y_i \geqslant y_{min}|z_i\} \qquad (2.8)$$

where the operator A transfers the set of possible components from the "zero" signals or "background" distinguishable signals of intensity greater than y_{min} in the position z_i to the set of possible components of the analyzed sample. The position z_i is a deterministic variable or at least the rank of the values z_i for individual components i $(i=1,2,\ldots,k)$ remains unchanged for the particular analytical procedure. The importance of the "position" z_i of a signal will be shown in a simple example of the method of distinguishing between Ag^+ and Pb^{2+} in a solution: a signal of intensity $y_i \geqslant y_{min}$ in the position z_i can be formed, for example, (1) by a white precipitate that arises from dilute HCl, (2) by the change in color that develops when carrying out a spot test by a chromate, or (3) by a spectral emission line. In case 1 the rise of the white precipitate by dilute HCl does not enable us to distinguish between Ag^+ and Pb^{2+}; distinguishing Ag^+ and Pb^{2+} would be possible only through determining other properties of the precipitate. However, in case 2 a different "position" in an imaginary coloring scale (i.e., different coloring) enables us to distinguish Ag^+ (a brown-red spot) from Pb^{2+} (a yellow spot). Spectral lines of Ag in case 3 at $\lambda=3280.68$ and 3382.90 Å enable us to distinguish Ag^+ from Pb^{2+} as does the fact that Pb has no spectral line in the spectral domain 3000 to 3500 Å. In qualitative analysis the a priori and a posteriori uncertainties, or the distributions $p_0(x)$ and $p(x)$, are characterized by the number of possible but so far unidentified components or by the number of possible combinations of the components.

The result of a *quantitative analysis* ξ_i is determined from the intensity η_i of the signal according to

$$E[\xi_i] = g_i(E[\eta_i]) \tag{2.9}$$

and the connection of $E[\xi_i]$ with X_i is obvious regarding the validity of the equality (2.1). If the dependence of the intensity η_i of the signal on the amount of the substance to be determined is known, for example, from the stoichiometry of the chemical reaction occurring in the determination, and if this dependence is expressed, for example, by the equivalent c_i, the mean content of the component to be determined is

$$\bar{x}_i = \frac{c_i}{m}\bar{y}_i \tag{2.10a}$$

where $\bar{y}_i$ is the mean intensity of the signal of the ith component obtained from one weighed amount m, or

$$\bar{x}_i = \frac{1}{n_p} \sum_{j=1}^{n_p} \frac{c_i}{m_j} y_{i,j} \tag{2.10b}$$

where the subscript j ($j = 1, 2, \ldots, n_p$) denotes parallel determinations. The values y_i and m or $y_{i,j}$ and m_j are those taken on by the random variable. Because the precision of weighing (i.e., determination of the weight m or m_j) is substantially greater than the precision of determination of the signal, we often take only the intensity η_i of the signal for a random variable that takes on the values $y_{i,j}$ for $j = 1, 2, \ldots, n_p$ for individual parallel determinations.

However, more frequently, we do not know the function g_i in the equality (2.9) and have to determine it by calibration (i.e., by measuring the values y_i for various values X_i and by graphically or numerically processing their dependence). For the determination of the content of the component being tested for in the analyzed sample,

$$\bar{x}_i = \frac{1}{m} g_i(\bar{y}_i) \qquad (2.11a)$$

or

$$\bar{x}_i = \frac{1}{n_p} \sum_{j=1}^{n_p} \frac{1}{m_j} g_i(y_{i,j}) \qquad (2.11b)$$

Considering that in the ideal case $E[\xi_i] = X_i$ should hold, the function g_i of the equalities (2.9), (2.11a), and (2.11b) should be inverse to the function f_i of (2.1). Here m_j and $y_{i,j}$ are values assumed by random variables and g_i is a nonstochastic function.

In quantitative analyses, the a priori uncertainty is characterized by a probability density $p_0(x)$ and most frequently by the uniform distribution in the interval of values within which we expect to find the content of the determined component. In some cases it can be characterized by the distribution of the results of a preceding analysis. The a posteriori uncertainty is always expressed by the probability distribution $p(x)$ of the results of parallel determinations.

As far as the scheme of the process of obtaining information about the quantitative composition of the analyzed sample is concerned, we have to distinguish between the specification of the scheme in Figure 2.1 for the case where we know the stoichiometric equivalent c_i and the case where we do not know it, and then determine by calibration the function g_i in (2.9) or the function f_i in (2.1) and take the function g_i as the inverse. In the former case (i.e., if we know the value c_i for various components $i = 1, 2, \ldots, k$) the system for decoding the intensities of the signals must be furnished by a "library" of the values c_i. It is arbitrary whether this library is realized in tables or in values stored in the memory of a computer or if we compute the values c_i from given atomic and molecular weights for a given stoichiometric ratio. This scheme is illustrated in Figure 2.5. However, if we determine the function g_i in an experimental way (i.e., by calibration), the

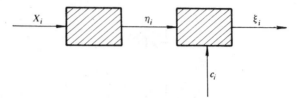

Figure 2.5 Scheme of quantitative analysis II. c_i = the stoichiometric equivalent.

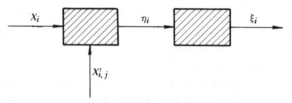

Figure 2.6 Scheme of quantitative analysis III. $X'_{i,j}$ refer to samples of known content.

process of obtaining information about the quantitative composition of an analyzed sample corresponds rather closely to the scheme illustrated in Figure 2.6. Here we have to insert a few reference samples with a known content of the determined component $X'_{i,j}$ ($j = 1, 2, \ldots, n$) into the first system. In terms of the values $X'_{i,j}$ the function g_i is determined; this is then used to compute x_i from the value η_i found for the unknown content X_i in the analyzed sample. Of course, the values $X'_{i,j}$ are known and we take them as fixed or at least more precisely known than other values we work with, so we do not need to consider them as random variables.

2.6 PROPERTIES OF ANALYTICAL RESULTS

The result of a quantitative analysis ξ_i is a random variable having a probability distribution with a describing function $p(x)$.

The difference between its expected value $E[\xi_i] = \mu_i$ and the true value of the ith component (i.e., $|\mu_i - X_i| = \delta_i$) characterizes the accuracy of the results. As a random variable the result of the analysis also has a certain variance $V[\xi_i]$, which expresses the precision of the results. Both properties, the accuracy and the precision, are mutually independent. The results can be precise (i.e., the agreement of individual results with the mean and among themselves can be very good), but at the same time the results need not be accurate and can be biased (i.e., subject to a certain systematic error). Another time the results can be less precise even though the mean of a larger number of parallel determinations can agree well with the true value. Results that are both precise and accurate are usually said to be *reliable*. A more exact examination of precision and accuracy will be given from the point of view of statistics in Chapter 5; here we will regard them as properties of the results of the analysis (i.e., properties of information).

Information is being defined as a decrease in the uncertainty of our knowledge of a random variable having a certain probability distribution. Uncertainty is always closely connected with a probability distribution, especially with its variance, in such a way that the greater the variance of the distribution of the results is (i.e., the less precise the results are), the greater is the uncertainty. The accuracy of the results (i.e., the difference $\delta_i = |\mu_i - X_i|$) can be favorably influenced by an appropriate calibration; when this is not possible and the results remain subject to a systematic error, it is possible to consider the noncentral uncertainty as an interval that is symmetrical about the true value X_i and not about the expected value. Since the noncentral uncertainty is always greater than the central uncertainty (which we use for the case $\delta_i = 0$), the quantitative side of the information is dependent on both the precision and the accuracy of the results. Details will be presented in Section 6.4, but it is obvious that reliable analytical results always represent the greatest gain in information.

2.7   TEMPORAL AND ECONOMIC VIEWS OF OBTAINING INFORMATION

In carrying out analyses we endeavor not only to obtain all relevant information, but we try to obtain this information in the shortest time and at the least cost. Therefore, in judging analytical methods we often take into account the duration of the analysis t_A and the cost τ_A required to carry out the analysis.

Sometimes it appears to be useful to divide these factors into one part independent of, and another part dependent on, the number of parallel determinations n_p; thus,

$$t_A = t_0 + n_p t_n \qquad (2.12a)$$

$$\tau_A = \tau_0 + n_p \tau_n \qquad (2.12b)$$

Time is usually considered in minutes or seconds and cost in currency units. It is obvious that we have to insert the time and costs for carrying out the analysis and evaluating the results into the figures for time t_A and costs τ_A. If we can express the financial equivalent of the time, it is sufficient to judge an analytical method only by using costs; from this point of view the factor τ_A appears to be more commonly adoptable than t_A. Then we have to insert such costs as wages of the analysts, rent for laboratories, time amortization of apparatus, and so on, into their financial equivalents, along with those of economic losses caused by waiting for results from the laboratory.

Temporal and economic aspects are involved in considering information performance and information profitability, which are important concepts when comparing and optimizing analytical methods and procedures.

References

1. H. Malissa, *Automation in und mit der analytischen Chemie*, Verlag der Wiener Medizinischen Akademie, Vienna, 1972.

2. H. Kaiser, *Z. Anal. Chem.* **260**, 252 (1972).

3. K. Doerffel, *Chem. Anal.* **17**, 615 (1972).

4. T. H. Apostol, *Calculus*, Vol. II, Blaisdell Publishing Company, Waltham, Mass., 1969.

5. H. Kaiser, *Anal. Chem.* **42**, No. 2, 24A; No. 4, 26A (1970).

6. K. Doerffel, *Z. Chem.* **8**, 236 (1968).

7. W. E. van der Linden, *Z. Anal. Chem.* **269**, 26 (1974).

8. H. Kaiser, *Z. Anal. Chem.* **209**, 1 (1965), **216**, 80 (1966).

9. J. D. Winefordner et al, *Anal. Chem.* **36**, 1939, 1947 (1964), **39**, 1495 (1967).

10. L. A. Currie, *Anal. Chem.* **40**, 586 (1968).

BASIC CONCEPTS OF PROBABILITY THEORY

3.1 INTRODUCTION TO PROBABILITY THEORY

3.1.1 Sample Space

The problems in probability theory are formulated with respect to a universal set S of all possible outcomes of a given experiment. Such a set will be called a *sample space* and its elements (i.e., individual results) will be *sample points* or *elementary events*.

Example. If we draw cards from a complete 52-card deck and are interested in their values, then $S = \{A, 2, 3, \ldots, 9, 10, J, Q, K\}$.

Example. The experiment of tossing a coin until a tail appears gives rise to a sample space with more than a finite number of elements, since $S = \{T, HT, HHT, HHHT, \ldots\}$ (H stands for heads and T for tails).

Example. If we toss a coin twice or toss two coins at the same time, either experiment provides the sample space $S = \{(HH), (HT), (TH), (TT)\}$ of ordered pairs. If we set $C = \{H, T\}$, then S is obviously the Cartesian product $S = C \times C$.

Together with a sample space S we will consider its subsets and call them *events*. We will denote by A_S the class of all subsets of S. It can be easily shown that a finite sample space S of n elements generates a class A_S of 2^n subsets, while a countable sample space S produces an uncountable class A_S.*

Example. For the space $S = \{H, T\}$ we have $A_S = \{\varnothing, \{H\}, \{T\}, \{H, T\}\}$, where $\varnothing$ stands for the empty set.

The examples introduced above have represented *discrete* sample spaces (i.e., such that are at most countable). However, many experiments yield nondiscrete sample spaces containing an uncountable number of sample points. This is the case, for example, whenever S is an interval of real numbers or a two-dimensional interval of pairs of real numbers. Then A_S is also uncountable. The class A_S always fulfills the properties of an algebra (respectively, of a σ-algebra).* We will consider uncountable sample spaces together with a σ-algebra of the events. Since the class A_S is also a σ-algebra, we will understand by the symbol A_S any σ-algebra.

In our next discussion we will deal with events rather than with elements or subsets, so the usual set arithmetic will be interpreted as follows.

If A_1 and A_2 are events in A_S, then:

$A_1 \cup A_2$ is the event that at least one of A_1 and A_2 occur.

$A_1 \cap A_2$ is the event that both A_1 and A_2 occur.

$A_1 \backslash A_2$ is the event that A_1 but not A_2 occurs.

A_1' is the event that A_1 does not occur.

$A_1 \subset A_2$ means that the occurrence of A_1 implies the occurrence of A_2.

S is the *sure event*, it must occur and for all $A \in A_S$, $A \subset S$.

$\varnothing$ is the *impossible event*; it never occurs.

---

*An infinite set is said to be *countable* if it can be put into one-to-one correspondence with the set of positive integers; otherwise, it is *uncountable*.

*An algebra (a σ-algebra) is a nonempty class $\mathcal{C}$ of the subsets of a set S for which the following hold:
1. The complement of each subset in $\mathcal{C}$ belongs to $\mathcal{C}$.
2. The union of any finite (countable) number of subsets in $\mathcal{C}$ belongs to $\mathcal{C}$.

Definition. The events A_1 and A_2 are *mutually exclusive* if they are disjoint [i.e., $A_1 \cap A_2 = \varnothing$ (they cannot occur simultaneously)].

3.1.2 Definition of Probability

Definition. *A set function* is a function the domain of which is a nonempty class of sets and that maps into the set of real numbers.

Example. Let $S = \{1,2,3,4,5,6\}$, for example, be the sample space of the outcomes on a die. Define a function g for all $A \in A_S$ so that $g(A)$ is the sum of the numbers in A. Then $g(A)$ is always real and g is a set function. We have, for example, $g(\varnothing) = 0$, and if $A = \{1,2,4\}$, then $g(A) = 7$.

Definition. A sample space S and a σ-algebra of its subsets are given. A set function P defined on A_S is called a *probability function* and $P(A)$ is called a *probability of the event A* if and only if:

(a) $P(A) \geqslant 0$ for all $A \in A_S$.
(b) $P(S) = 1$.
(c)   If A and B are mutually exclusive events, then $P(A \cup B) = P(A) + P(B)$.
(d)   If $A_1, A_2, A_3, \ldots$ is a sequence of mutually exclusive events, then $P(\cup_i A_i) = \Sigma_i P(A_i)$.

Axioms (c) and (d) require that $\{A \cup B\}$ and $\cup_i A_i$, respectively, belong to A_S. However, this is guaranteed by A_S being an algebra or a σ-algebra. Axiom (d) is redundant in the case of finite spaces. Axiom (c) can be extended to a finite number of mutually exclusive events $A_1, A_2, \ldots, A_n$.

Note that we started with a sample space and then defined a set function on the class of events A_S, which is a probability function if it fulfills the requirements of the definition. This differs, of course, from our earlier experience with the functions in mathematical analysis. There we started with mapping and then investigated the domain and the set of the images (function values). In our work with probabilities we will start with the domain and the set of the images and only then will we create a mapping.

The search for a method of forming such a mapping is simple if we deal with a finite or infinitely countable sample space. Thus, if $S = \{s_1, s_2, s_3, \dots\}$ and P is a probability function on A_S, $P(\{s_i\})$ is defined for all elementary events $\{s_i\} \in A_S$. For the sake of simplicity we will write $P(s_i)$ instead of $P(\{s_i\})$. Furthermore, if $A \in A_S$ consists of a finite or countable number of elementary events (and $A \neq \varnothing$), then $A = \{s_{a_1}, s_{a_2}, s_{a_3}, \dots\}$, $s_{a_i} \in S$, and we can express A as a disjoint union of its elementary events:

$$A = \bigcup_i \{s_{a_i}\} = \bigcup_{s_j \in A} \{s_j\}$$

(Obviously, each $s_{a_i} \in A$ is an $s_j \in S$, and if we unite elementary events in A, we unite those elementary events in S that belong to A.) Since P is a probability function, we get

$$P(A) = P\left(\bigcup_{s_i \in A} \{s_i\} \right) = \sum_{s_i \in A} P(s_i)$$

and thus have expressed the probability of an event as the sum of the probabilities of the sample points that form the event.

So a question has arisen: why not proceed in the inverse way; that is, why not define P explicitly only for the elementary events and then set $P(A) = \sum_{s_i \in A} P(s_i)$ for all $A \in A_S$? The following theorem gives a positive answer under obvious limiting assumptions.

Theorem. Let $S = \{s_1, s_2, s_3, \dots\}$ be a finite or infinitely countable sample space. Define a set function P on A_S such that:

(a) $P(s_i) \geqslant 0$ for all sample points $\{s_i\} \in A_S$.
(b) $\sum_{s_i \in S} P(s_i) ` 1$.
(c) $P(A) = \sum_{s_i \in A} P(s_i)$ for $A \in A_S$.
Then P is a probability function. (We omit the proof.)

Example. An experiment consists of trials in which an event E either occurs or does not occur. The trials are repeated until the event E occurs. Now $S = \{E, E'E, E'E'E, \ldots\}$. Let s_i denote the elementary event in S when we need i trials: for example, $s_4 = E'E'E'E$. Let us set $P(s_i) = \frac{1}{6} \cdot (\frac{5}{6})^{i-1}$ for $\{s_i\} \in A_S$ and require that $P(A) = \Sigma_{s_i \in A} P(s_i)$ for $A \in A_S$ [if, e.g., A is an event needing more than three trials, then $P(A) = \Sigma_{i=4}^{\infty} \frac{1}{6} \cdot (\frac{5}{6})^{i-1}$, because $A = \{s_4, s_5, \ldots\}$]. We ask whether P is a probability function. Since

$$\sum_{s_i \in S} P(s_i) = \sum_{i=1}^{\infty} \frac{1}{6} \cdot \left(\frac{5}{6}\right)^{i-1} = \frac{\frac{1}{6}}{1 - \frac{5}{6}} = 1$$

the answer is positive.

Our considerations of the probability have thus concentrated around three fundamental concepts: the sample space S, the class A_S of the events in S (a σ-algebra), and the probability function P.

Definition. An ordered triplet $\{S, A_S, P\}$ is called a *probability space*.

Any problem concerning probabilities is inserted in a probability space; before this space is described, no problem can be solved.

One sample space can be transferred into a probability space by various probability functions. If we have no experimental evidence, we take into account the factors that influence the occurrence of the events and select a probability function that most closely maps the physical world as we know it. In such cases we speak of an *a priori probability*.

Often, a probability function is chosen according to the following theorem, which is a consequence of the definition of probability functions in finite sample spaces.

Theorem. Denote by $n(A)$ the number of elementary events composing an event A. A finite sample space $S = \{s_1, s_2, \ldots, s_N\}$ is given. If we define a set function on A_S in such a way that

$$P(A) = \frac{n(A)}{N}$$

for $A \in A_S$, then P is a probability function. It follows that

$$P(s_i) = \frac{1}{N}$$

for $i = 1, 2, \ldots, N$. The constructed probability space is called *uniform*.

As an alternative, we can determine a probability function by repeating an experiment several times, observing the relative frequencies of its possible outcomes (in a finite sample space), and using them for proposing an approximate probability function. We can so proceed, for example, in certain scientific hypotheses. We speak then about setting an *a posteriori probability function*.

The structures of the sample spaces were not complex in the preceding cases and it was easy to find out if a sample point belonged to an event. The matter is not so simple once we proceed to uncountably infinite sample spaces. We encounter some difficulties if we attempt to define a σ-algebra on the real axis containing all intervals.

We now ask how we can introduce a probability function. When the sample space was finite or countable, we saw that it was sufficient to determine the probability for each sample point. The probabilities of other events could afterward be calculated so that we considered the events as a discrete class of elementary events and adopted axiom (c) or (d). We can proceed similarly when the sample space is uncountable. If the domain is given by a σ-algebra, we need not present probabilities for all its sets. A probability measure is well specified on the real axis when we assign numbers to intervals $\{x; x \leqslant t\}$ for every real number t. If these numbers are known, probabilities for other sets of the σ-algebra can be calculated. We will deal more with them in the section on distribution functions.

3.1.3 Properties of Probability

Some simple properties of probability functions follow directly from their definition, and since they are often used in counting probabilities, we will summarize them here. We assume for the probability functions that they are defined on a σ-algebra so that all events occurring in the following theorems belong to the domain of the function P.

If A is any event in A_S, the following hold:

Theorem 1. For the complementary event A',

$$P(A')=1-P(A)$$

Proof. Obviously, $A \cup A' = S$, $A \cap A' = \emptyset$. Using axioms (b) and (c) in the definition of probability we get

$$P(S)=P(A \cup A')=P(A)+P(A')=1$$

hence

$$P(A')=1-P(A)$$

Theorem 2. The probability of the impossible event is zero.

Proof. We use Theorem 1 for $S'=\emptyset$:

$$P(\emptyset)=1-P(S)=0$$

Theorem 3. If $B \subset A$, holds for events $A \in A_S$ and $B \in A_S$,

then $P(B) \leqslant P(A)$ and $P(A \backslash B)=P(A)-P(B)$

Proof. According to the assumption, we can write

$$A = B \cup (A \backslash B)$$

whereby the events on the right-hand side are mutually exclusive.

Thus, we have, from axiom (c),

$$P(A) = P(B) + P(A \setminus B)$$

from where both conclusions follow [the first one needs, in addition, the use of the property $P(A \setminus B) \geqslant 0$].

Theorem 4. For any two events $A \in A_S$ and $B \in A_S$,

$$P(A \cup B) = P(A) + P(B) - P(A \cap B)$$

Proof. The union $A \cup B$ can be written as the union of two exclusive events:

$$A \cup B = A \cup [B \setminus (A \cap B)]$$

Using axiom (c), we have

$$P(A \cup B) = P(A) + P(B \setminus (A \cap B))$$

and by Theorem 3,

$$P(A \cup B) = P(A) + P(B) - P(A \cap B)$$

Theorem 5. For any three events A, B, and C of A_S,

$$P(A \cup B \cup C) = P(A) + P(B) + P(C) - P(A \cap B) - P(A \cap C) \\ - P(B \cap C) + P(A \cap B \cap C)$$

Proof. It follows immediately if we adopt the preceding theorem twice and use the equality

$$A \cap (B \cup C) = (A \cap B) \cup (A \cap C)$$

We need to remark that for any event $A \in A_S$,

$$0 \leqslant P(A) \leqslant 1$$

The first inequality is required by axiom (a); the second one is true because S belongs to A_S and for any $A \in A_S$, $A \subset S$, and we can use Theorem 3 and axiom (b).

3.1.4 Conditional Probability

Definition. Given a probability space $\{S, A_S, P\}$, A and B are events in A_S and $P(A) > 0$. The set function that assigns the number

$$P(B|A) = \frac{P(A \cap B)}{P(A)} \tag{3.1}$$

to the event B is called the *conditional probability function* given event A and $P(B|A)$ is the *conditional probability* of the event B given event A.

It can be verified that the new function fulfills the axioms of the probability measure for $A \in A_S$ such that $P(A) > 0$. The term "given A" distinguishes this function from ordinary probability functions. Here all outcomes of a random experiment correspond to sample points in A (i.e., it is given that the event A occurred).

It follows from the definition that

$$P(A \cap B) = P(A) \cdot P(B|A) \tag{3.2}$$

For three events A, B, and C in A_S with $P(A) > 0$ and $P(A \cap B) > 0$,

$$P(A \cap B \cap C) = P(A) \cdot P(B|A) \cdot P(C|(A \cap B)) \tag{3.3}$$

The proof is carried out by repeated use of the definition of the conditional probability.

3.1.5 Stochastic Independence

Definition. Given a probability space $\{S, A_S, P\}$, two events A and B are *stochastically independent* if

$$P(A \cap B) = P(A) \cdot P(B) \tag{3.4}$$

The most outstanding feature of this definition is that the stochastic independence is an expression of equality between two numbers. A decision on stochastic independence is difficult without the probability function. It requires intuitive feeling for its interpretation. The following property relates the concept of stochastic independence to the idea of the conditional probability function and is the first step in developing the necessary intuition.

Corollary. If A and B are stochastically independent, then

$$P(A|B) = P(A)$$
$$P(B|A) = P(B)$$

whenever $P(A) > 0$ and $P(B) > 0$.

Proof. It follows from the definition of the conditional probability that

$$P(A|B) = \frac{P(A \cap B)}{P(B)} = \frac{P(A) \cdot P(B)}{P(B)} = P(A)$$

and similarly for the second part.

Inversely, from equalities $P(A|B) = P(A)$ and $P(B|A) = P(B)$ follows the stochastic independence of A and B.

Thus, if A and B are stochastically independent events, the knowledge that one of the events has occurred does not change the probability of the occurrence of the other event. It is just this sense in which both events are stochastically independent. When the conditions of an experiment guarantee that the frequency of occurrence of one event does not influence the occurrence of the other event, we can assume the events to be stochastically independent.

Remark. The term "statistical independence" means the same as stochastic independence. We will often simply use the term "independence."

It remains to extend the concept of stochastic independence to more than two events.

Definition. Given a probability space $\{S, A_S, P\}$, the events $A_1, A_2, \ldots, A_n$ are called *stochastically independent* if

$$P(A_i \cap A_j) = P(A_i) \cdot P(A_j) \qquad \text{whenever} \quad i \neq j$$

$$P(A_i \cap A_j \cap A_k) = P(A_i) \cdot P(A_j) \cdot P(A_k) \qquad \text{whenever} \quad i \neq j \neq k$$

and

$$P\left(\bigcap_{i=1}^{n} A_i\right) = \prod_{i=1}^{n} P(A_i)$$

The first row represents $\binom{n}{2}$ equalities, the second one $\binom{n}{3}$ equalities, and so on. The first $\binom{n}{2}$ equalities assure *pairwise independence*.

3.2 MATHEMATICAL MODEL

A sample space and a probability function represent a reasonable mathematical model for a random experiment, although dealing with it is often uncomfortable. A list of sample points belonging to an event may cause difficulties and a probability function cannot be easily manipulated, for it is a set function. Therefore, we will introduce an alternative mathematical model working with point functions so that ordinary calculus may be used in probability theory. This model consists of two units: a random variable and a describing function. A random variable substitutes for the role of a sample space by relating possible outcomes of a random experiment to real numbers. A describing function serves for calculating probabilities.

3.2.1 Random Variables

The random variable is a mathematical mechanism for relating outcomes of random experiments to real numbers.

Definition. Let $\{S, A_S, P\}$ be a probability space. Let $\xi(s)$ be a real function defined at every sample point $s \in S$ such that for every real number x, the event E_x, for which $\xi(s) \leqslant x$, belongs to A_S. We will call the function $\xi(s)$ a *random variable* relative to A_S.

Thus, the probability space $\{S, A_S, P\}$ associated with the basic sample space S can be used to define a sample space $\{\mathfrak{R}_1, \mathfrak{B}_1, P_\xi\}$ associated with the real line $\mathfrak{R}_1$ by setting $P_\xi(A') = P(\xi^{-1}(A'))$ for every set $A' \in \mathfrak{B}_1$ (a σ-algebra on $\mathfrak{R}_1$), where $A = \xi^{-1}(A') \in A_S$ consists of all sample points in S for which $\xi(s) \in A'$. We say that the triple $\{S, A_S, P\}$ induces the probability space by the means of the random variable $\xi(s)$. Many problems in probability can be answered more readily in this induced space rather than in the original one.

The random variable does nothing more or less than map the sample space onto a set of real numbers, or label each outcome of the random experiment with a real number.

The word "variable" in the name "random variable" is misleading since random variable is a function. The designation "random variable" is to be viewed as a unit and not as a variable that happens to be a random (whatever that means). This notational inconvenience is firmly anchored in the language of probability theory.

Some random experiments have multidimensional outcomes. Then one random variable is defined for each dimension and a vector is formed. If, for the given probability space $\{S, A_S, P\}$, $\xi_1(s), \xi_2(s), \ldots, \xi_k(s)$ are k random variables, they form an induced probability space $\{\mathfrak{R}_k, \mathfrak{B}_k, P_{\xi_1, \ldots, \xi_k}\}$, where $\mathfrak{R}_k$ is Euclidean k-dimensional space. It is convenient to refer to $(\xi_1(s), \xi_2(s), \ldots, \xi_k(s))$ as a *k-dimensional* or *vector random variable*.

3.2.2 Describing Functions

From now on we shall, for the sake of simplicity, omit the argument $s \in S$ in dealing with random variables. If ξ is a random variable we will often use the notation $P\{\xi \in A'\}$ instead of

$P_\xi(A')$ for any $A' \in \mathcal{B}_1$. Specially, if A' is an interval $(a,b\rangle$, we will understand that the probability $P\{\xi \in A'\}$ can be written $P\{a < \xi \leqslant b\}$; similarly, for a one-point set $\{x\}$ we will write $P\{\xi = x\}$. Analogously, we will denote probabilities of a k-dimensional random variable $(\xi_1, \xi_2, \ldots, \xi_k)$.

A describing function forms the second part of our new mathematical model. It is a common name for functions, enabling calculation of probabilities. We will examine them briefly, since they represent tools that we will use frequently in the following chapters.

The Cumulative Distribution Function

Definition. Given a probability space $\{S, A_S, P\}$ and a random variable ξ defined on S, the *cumulative distribution function* of the random variable ξ is the point function $F_\xi(x)$ defined by

$$F_\xi(x) = P\{\xi \leqslant x\} \tag{3.5}$$

whose domain is the set of real numbers $\mathcal{R}_1$.

Remark. The subscript on F will be omitted whenever the random variable it is concerned with is obvious. We will also leave out the adjective "cumulative". The argument x stands for any point in the domain of the function. Any other letter can be used to represent the argument. We will denote this variable mostly by a letter from the Latin alphabet corresponding to the Greek letter chosen for the denotation of the random variable (e.g., the values of random variables ξ, η, and ζ will bear the letters x, y, and z).

The event $\{s; \xi(s) \leqslant x\}$ is for any real number x the set of all sample points that are mapped by the random variable ξ into numbers less or equal to x. The definition of the random variable assures that all such events are in the domain of the probability function.

Every distribution function has the following properties:

1. $\lim_{x\to\infty} F(x) = F(+\infty) = 1, \lim_{x\to-\infty} F(x) = F(-\infty) = 0.$
2. $F(x)$ is a monotone, nondecreasing function.
3. $F(x)$ is everywhere continuous from the right; that is,

$$\lim_{x\to x_0+} F(x) = F(x_0 + 0) = F(x_0)$$

These properties follow from the definition and properties of probability functions.

We are now interested in how we can calculate probabilities for other subsets of $\mathcal{R}_1$ than those of the type $\{\xi \leqslant x\}$. They are given by the following properties:

1. $P\{x_1 < \xi \leqslant x_2\} = F(x_2) - F(x_1)$ for any x_1 and x_2, $x_1 < x_2$.
2. $P\{\xi > x\} = 1 - F(x).$
3. $P\{\xi = x\} = F(x) - F(x - 0).$

If $F(x)$ is continuous at the point x, the last property shows that $P\{\xi = x\} = 0$, since $F(x - 0) = F(x)$. If x is a jump point of $F(x)$, then $F(x - 0) \neq F(x)$, so $P\{\xi = x\} > 0$ and represents the size of the jump at the point x.

Examples of distribution functions are shown in Figure 3.1. Since the properties of random variables with distribution functions continuous over the entire real line seem to be different from those of the ones whose distribution functions contain jump points, the distribution function will obviously provide a natural means of distinguishing random variables. We will therefore classify random variables into two types.

Discrete Random Variables

Definition. A random variable is *discrete* if its distribution function is a step function.*

---

*A point function is called a *step function* if it varies its values in at most a countable number of points $x_1, x_2, x_3, \ldots$ not having a finite limit point.

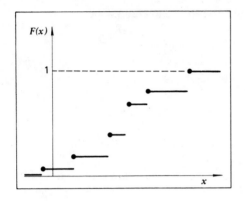

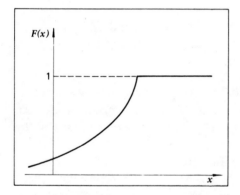

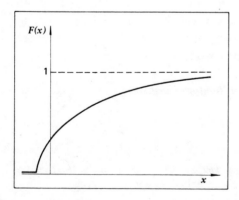

Figure 3.1 Examples of distribution functions.

40

It is obvious that on discrete sample spaces containing a finite or countable number of sample points only discrete random variables can be defined.

For the distribution function of a discrete random variable having jumps of size $p(x_i) = P\{\xi = x_i\}$ at the points x_i,

$$F(x) = P\{\xi \leqslant x\} = \sum_{\substack{i \\ x_i \leqslant x}} p(x_i)$$

where we sum over x_i less than or equal to x. If $x \to \infty$, we get

$$F(+\infty) = \sum_i p(x_i) = 1$$

Continuous Random Variables

Definition. A random variable is *continuous* if its distribution function is continuous in $\mathfrak{R}_1$ and has a derivative everywhere except, possibly, at a finite number of points.

As we have seen, continuous random variables have zero probabilities at all their values. However, intervals may have nonzero probabilities. This follows from the nature of the sample space: a continuum is a useful mathematical abstraction allowing the advantages of the calculus, although it can deceive our intuition. It is just this class of random variables that plays the most important role in applications to analytical chemistry.

Other Describing Functions

A random variable and its distribution function form a perfect mathematical model of a random experiment from which probabilities can be computed. However, the mathematical form of distribution functions is too unwieldy to easily compute probabilities in many cases. So we are led to introduce other describing functions, which are the most important ones from a practical point of view.

Definition. Let ξ be a discrete random variable taking on the values $x_1, x_2, x_3, \ldots$. The set of all sample points for which ξ takes on a fixed value x_i forms an event $\{s; \xi(s) = x_i\}$. We will call the function

$$P\{s; \xi(s) = x_i\} = p(x_i) \tag{3.6}$$

the *density function* of the random variable ξ.

If we leave out the subscript i, we can say that the density function $p(x)$ fulfills

$$p(x_i) \geqslant 0 \qquad \text{for all } i$$

$$\sum_i p(x_i) = 1$$

$$p(x) = 0 \qquad \text{for other } x \in \mathcal{R}_1$$

Since both the density function and the distribution function are describing functions, any of them can be derived from the other one. The following relations hold between them:

1. $p(x_i) = F(x_i) - F(x_i - 0)$.
2. $F(x) = \displaystyle\sum_{\substack{i \\ x_i \leqslant x}} p(x_i)$.

The relations are illustrated in Figure 3.2.

Definition. Let ξ be a continuous random variable. The function

$$f(x) = \frac{dF(x)}{dx} \tag{3.7}$$

will be called the *probability density* or the *frequency function*.

By definition of a continuous random variable, the derivative of $F(x)$ exists everywhere except, perhaps, at no more than a countable number of points. At those points where the derivative does

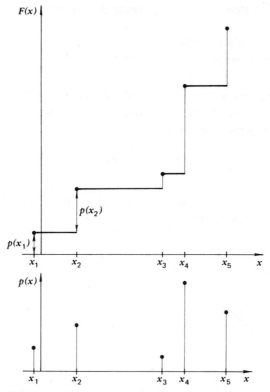

Figure 3.2 Distribution function and density function of a discrete random variable.

not exist, $f(x)$ is not formally defined, although appropriate values may be assigned. This causes no inconsistencies because $P\{\xi=x\}$ $=0$ for continuous random variables.

It follows from the definition of a derivative for a small $\Delta x>0$:

$$f(x)\Delta x \doteq F(x+\Delta x)-F(x)=P\{x<\xi\leqslant x+\Delta x\}$$

The number $f(x)$ itself is not a probability but is approximately proportional to a probability. In the differential form, the expression $f(x)\,dx$ is called a *probability element*.

Probability densities have following properties:

1. $f(x) \geqslant 0$ for all $x \in \mathcal{R}_1$.
2. $\int_{-\infty}^{\infty} f(x) dx = 1$.

The relation between the probability density and the distribution function is given by the definition of the probability density. Inversely, this relation can be written

$$F(x) = \int_{-\infty}^{\infty} f(t) dt \qquad (3.8)$$

The integral exists because $f(x)$ is defined almost everywhere. The relation is shown in Figure 3.3.

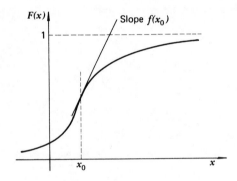

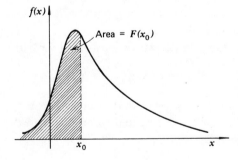

Figure 3.3 Distribution function and frequency function of a continuous random variable.

Now that the relations between $p(x)$ and $F(x)$ and between $f(x)$ and $F(x)$ have been established, the technique for calculating probabilities from $p(x)$ and $f(x)$, respectively, can be stated.

If ξ is a discrete random variable, we have

$$P\{a<\xi\leqslant b\} = \sum_{\substack{i \\ a<x_i\leqslant b}} p(x_i) \qquad (3.9)$$

and if ξ is a continuous random variable, then

$$P\{a<\xi\leqslant b\} = \int_a^b f(x)\,dx \qquad (3.10)$$

Both equations show that these functions make computing probabilities extremely simple. In the discrete case, the nonzero values of the density function are added, while for continuous random variables the area under the frequency function is computed.

3.2.3 Two-Dimensional Describing Functions

During investigations of random phenomena, the interplay between two or more random variables is often of primary interest. Handling a number of random variables simultaneously requires an extension of the describing functions introduced in the previous section. We have to account for the relative frequencies with which the random variables jointly take on their values.

We will consider only two-dimensional describing functions, as the extension to more dimensions is easy. We will start with the definition of the joint distribution function and explain its relation to one-dimensional distribution functions.

Definition. Given is a probability space $\{S, A_S, P\}$ and a pair of random variables (ξ, η) defined on S. The *joint (cumulative) distribution function* for the pair (ξ, η) is the two-dimensional point function

$$F_{\xi,\eta}(x,y) = P\{\xi \leqslant x, \eta \leqslant y\} \qquad (3.11)$$

defined on the real plane $\Re_2$.

Remark. $P\{\xi \leqslant x, \eta \leqslant y\}$ is the probability of the simultaneous occurrence (intersection) of the events $\{s; \xi(s) \leqslant x\}$ and $\{s; \eta(s) \leqslant y\}$. The subscripts at F refer to the random variables but will often be omitted if confusion will not result.

A joint distribution function always fulfills these properties:

1. $\lim_{x \to -\infty} F(x,y) = F(-\infty, y) = 0$,
   $\lim_{y \to -\infty} F(x,y) = = F(x, -\infty) = 0$.
2. $\lim_{x \to \infty, y \to \infty} F(x,y) = F(\infty, \infty) = 1$.
3. For any $h \geqslant 0$ and $k \geqslant 0$,
   $\Delta^2 F(x,y) = F(x+h, y+k) - F(x+h, y) - F(x, y+k) + F(x,y) \geqslant 0$
4. $F(x,y)$ is continuous from the right in each variable; that is,

$$F(x+0,y) = F(x, y+0) = F(x,y)$$

Figure 3.4 relates to the property 3. $\Delta^2 F(x,y)$ is the probability contained in the blank rectangle (along with the upper and right boundaries).

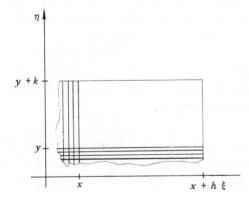

Figure 3.4   Domain of a two-dimensional distribution function.

When the joint distribution function for a pair of random variables is known, the individual distribution functions, called *marginal* distribution functions, can be determined as follows.

$$F_\xi(x) = F_{\xi,\eta}(x, \infty)$$
$$F_\eta(y) = F_{\xi,\eta}(\infty, y)$$

$F_{\xi,\eta}(x, \infty)$ determines the probability $P\{\xi \leqslant x\}$. We can easily verify that $F_\xi(x)$ and $F_\eta(y)$ satisfy the conditions of one-dimensional distribution functions.

The mathematical form of a joint distribution function depends on the classification of the random variables. If both are discrete, the joint distribution function can be thought of as a two-dimensional staircase going up in either positive direction in discrete steps. When both are continuous, the joint distribution function represents a surface.

Again, we can simplify computing probabilities by defining more useful describing functions. With regard to the predominating importance of continuous probability distributions in applications to analytical chemistry, we will restrict ourselves to treating only two-dimensional continuous random variables.

Definition. The *continuous* two-dimensional random variable (ξ, η) has a continuous joint distribution function $F(x, y)$ defined on $\mathcal{R}_2$ for which $\partial^2 F(x,y)/\partial x \partial y$ exists everywhere except, possibly, at no more than a countable number of points. The nonnegative function

$$f(x,y) = \frac{\partial^2 F(x,y)}{\partial x \partial y} \tag{3.12}$$

is the *joint probability density* of the pair (ξ, η) and

$$F(x,y) = \int_{-\infty}^{x} \int_{-\infty}^{y} f(u,v) \, du \, dv \tag{3.13}$$

Obviously,

$$\int_{-\infty}^{\infty}\int_{-\infty}^{\infty} f(x,y)\,dx\,dy = 1$$

For an arbitrary set $E \in \mathfrak{R}_2$, we get

$$P\{(\xi,\eta)\in E\} = \int_E\int f(u,v)\,du\,dv$$

Specifically, we have for the marginal distribution function of the random variable ξ,

$$F_\xi(x) = P\{\xi \leqslant x\} = \int_{-\infty}^{x}\int_{-\infty}^{\infty} f(u,y)\,du\,dy$$

If we take the integral for a twofold one, then

$$f_\xi(x) = \int_{-\infty}^{\infty} f(x,y)\,dy \qquad (3.14)$$

is the *marginal probability density* of the random variable ξ. Similarly, we can get

$$f_\eta(y) = \int_{-\infty}^{\infty} f(x,y)\,dx \qquad (3.15)$$

Either of the two describing functions defined above is sufficient to specify the joint distribution of a pair of random variables, since we can obtain one from the other. A pertinent question is how we can establish a joint describing function.

The problem of constructing such a function from the data belongs to statistics. The probability theory is concerned with hypothetical rules to approximate actual physical situations. It is left to the cleverness and experience of the person concerned to translate the ground rules into a describing function. There are a few techniques available for approaching this problem methodically. Stochastic independence leads to such a technique. The concept of stochastic independence fulfills a need similar to that satisfied by the notion of independent events.

Definition. The random variables ξ and η are *stochastically independent* if for all pairs $(x,y) \in \mathcal{R}_2$,

$$F_{\xi,\eta}(x,y) = F_\xi(x) \cdot F_\eta(y) \tag{3.15a}$$

Equivalently, ξ and η are stochastically independent if their joint probability density equals the product of the marginal probability densities for all $(x,y) \in \mathcal{R}_2$.

When two random variables are stochastically independent, their joint describing functions can be obtained from marginal describing functions without any additional information.

3.2.4 Expectations and Moments

The mathematical model that we developed for the description of a random experiment is satisfactory to compute probabilities. However, in addition to this, for any probability distribution some numbers can be defined in which information about the nature of the distribution is condensed and which enable us to extend the understanding of random phenomena.

At first we will explain the notation for the expectation operator, which is the basis for all the ideas in this section.

Definition. Let $g(x)$ be a real function of a real variable and ξ a random variable. The *expectation*, or *expected value*, of the function g of the random variable ξ is the number:

(a) In the discrete case,

$$E[g(\xi)] = \sum_i g(x_i) p_\xi(x_i)$$

if the series converges absolutely.

(b)   In the continuous case,

$$E[\,g(\xi)\,] = \int_{-\infty}^{\infty} g(x)f_{\xi}(x)\,dx$$

if the integral converges absolutely.

The expectation operator isolates the information about a probability distribution by defining certain numbers which are representative of the distribution. We will assume that all numbers defined by the means of the expectation operator exist (i.e., they are finite).

Definition.   The *moment of the* rth *order* $(r > 0,$ integer) of a random variable ξ is the number

$$\mu_r' = E[\,\xi^r\,] = \begin{cases} \sum_i x_i^r p(x_i) \\ \int_{-\infty}^{\infty} x^r f(x)\,dx \end{cases} \qquad (3.16)$$

The definition applies the introduced operator to the function $g(\xi) = \xi^r$. For $r = 1$ we get the *expected value* or *mean value* of the random variable ξ which will often be denoted by μ:

$$\mu_1' = E[\,\xi\,] = \mu = \begin{cases} \sum_i x_i p(x_i) \\ \int_{-\infty}^{\infty} x f(x)\,dx \end{cases} \qquad (3.17)$$

One way of highlighting the meaning of mean values is by analogy with the mass. For instance, if ξ is a discrete variable with mass points at $x_1, x_2, \ldots, x_n$, its mean value is a weighted average: the numerical value of each point on the real line is multiplied by its mass and all such products are summed.

Definition. The *central moment of the r*th *order* $(r > 0$, integer) of a random variable ξ is the number

$$\mu_r = E\left[(\xi - \mu)^r\right] = \begin{cases} \sum_i (x_i - \mu)^r p(x_i) \\ \int_{-\infty}^{\infty} (x - \mu)^r f(x)\, dx \end{cases} \quad (3.18)$$

The most important of these moments is that of second order and is called the *variance* and will be denoted $V[\xi]$ or σ^2. Hence,

$$\mu_2 = V[\xi] = \sigma^2 = E\left[(\xi - \mu)^2\right] = \begin{cases} \sum_i (x_i - \mu)^2 p(x_i) \\ \int_{-\infty}^{\infty} (x - \mu)^2 f(x)\, dx \end{cases} \quad (3.19)$$

The quantity σ, the positive square root of σ^2, is called the *standard deviation* of ξ.

The smaller the variance and the standard deviation of a random variable, the greater is the concentration of the probabilities around the mean value. This concentration can be labeled "dispersion about the mean value." The greater the concentration, the less the dispersion. The variance serves as a measure of the dispersion and exhibits different information from that of the expected value.

The most suitable way to calculate the variance is to use the following expression in moments:

$$\sigma^2 = E\left[\xi^2\right] - \mu^2 = \mu_2' - \mu^2 \quad (3.20)$$

This formula can be easily found by expanding the right-hand sides in (3.19).

Expectations have some important properties, which we summarize here. All expected values are assumed to exist.

1. If a and b are real numbers and ξ is a random variable, then

$$E[a\xi + b] = aE[\xi] + b$$

2. For arbitrary random variables $\xi_1, \xi_2, \ldots, \xi_n$

$$E\left[\sum_{i=1}^{n} \xi_i\right] = \sum_{i=1}^{n} E[\xi_i]$$

3. If $\xi_1, \xi_2, \ldots, \xi_n$ are stochastically independent random variables, then

$$E\left[\prod_{i=1}^{n} \xi_i\right] = \prod_{i=1}^{n} E[\xi_i]$$

Similarly, the variance has properties which are useful in both theoretical derivations and practical calculations.

1. If a and b are real numbers and ξ is a random variable, then

$$V[a\xi + b] = a^2 V[\xi]$$

2. If $\xi_1, \xi_2, \ldots, \xi_n$ are pairwise-independent random variables, then

$$V\left[\sum_{i=1}^{n} \xi_i\right] = \sum_{i=1}^{n} V[\xi_i]$$

Example. Let us calculate the variance of the difference of two independent random variables, ξ and η. According to the previous properties, we have

$$V[\xi - \eta] = V[\xi + (-\eta)] = V[\xi] + V[-\eta]$$
$$= V[\xi] + (-1)^2 V[\eta] = V[\xi] + V[\eta]$$

Thus, the result is the *sum* of the variances of the two variables.

Example. If a random variable ξ has expected value and variance, we will call

$$\zeta = \frac{\xi - E[\xi]}{\sqrt{V[\xi]}}$$

the *standardized random variable*. It follows that (using previous properties)

$$E[\zeta] = E\left[\frac{\xi - E[\xi]}{\sqrt{V[\xi]}}\right] = \frac{1}{\sqrt{V[\xi]}} E[\xi - E[\xi]] = 0$$

$$V[\zeta] = V\left[\frac{\xi - E[\xi]}{\sqrt{V[\xi]}}\right] = \frac{1}{V[\xi]} V[\xi - E[\xi]]$$

$$= \frac{1}{V[\xi]} V[\xi] = 1$$

3.3 SOME SPECIFIC PROBABILITY DISTRIBUTIONS

A number of random experiments can often be characterized by essentially the same mathematical model if the ground rules of the experiment are reduced to common terms. Although the values of certain parameters vary from one situation to the other, the basic structure of the model and the forms of the describing functions remain the same. Such models are important in applications of probability theory and it is necessary to get acquainted with them.

The statement that a random variable has a certain probability distribution means that its describing function has a prescribed mathematical form. Knowledge of the distribution or the assumption of it lies in the grounds of statistical inference, of which we will speak in another chapter. The origin of some probability distributions can be explained by a suitable physical experiment.

In this section we will examine those distributions that appear to be needed in applications to analytical chemistry. Two discrete distributions will be investigated first. Afterward, three continuous distributions that will be encountered repeatedly in later chapters will be discussed.

3.3.1   The Binomial Distribution

A conceptual random experiment generating the binomial distribution can be described in general terms as follows. A certain trial is repeated n times. In each trial the presence or absence of a certain elementary event is noted. If the event occurs, the trial is said to result in success; if it is absent, the trial ends in failure.

The sample space S for this random experiment can be defined by

$$S = \{ k_i; i = 1, 2,, \ldots, n; k_i = 0 \text{ or } 1 \}$$

A digit 1 means that the trial resulted in a success while a digit 0 implies failure and each sample point is labeled with n binary digits, the ith digit denoting the outcome of the ith trial.

We assume that the probability of success in an individual trial is the same for all trials. Furthermore, we require that the outcome of one trial be *independent* of the outcome of any other trial.

Let us define a random variable ξ on S as the number of successes in the n trials. If θ is the probability of success in any one trial and s' is a sample point labeled with k unities and $(n-k)$ zeros,

$$P(s') = \theta^k (1-\theta)^{n-k}$$

The event $\{s; \xi(s) = k\}$ contains $\binom{n}{k}$ sample points having k unities and $(n-k)$ zeros. Hence, the density function of ξ can be written

$$P_n\{\xi = k\} = p(k) = \binom{n}{k} \theta^k (1-\theta)^{n-k} \qquad k = 0, 1, 2,, \ldots, n$$

$$(3.21)$$

Obviously, $p(k) > 0$. Since these probabilities appear to be members of the binomial expansion of $[\theta + (1-\theta)]^n = 1$, the distribution of ξ is called *binomial*. The combinatorial coefficient is

$$\binom{n}{k} = \frac{n!}{k!(n-k)!}$$

The distribution function of the binomial random variable is given by

$$F_\xi(k) = \sum_i p(i) = \sum_i \binom{n}{i} \theta^i (1-\theta)^{n-i}$$

where we sum up over all nonnegative integers less than or equal to k.

The binomial distribution has two parameters, the number of independent trials n and the probability θ. In practical applications, n is known from the rules of the experiment, while θ is usually estimated from observations by statistical methods. Examples of binomial density functions are illustrated in Figure 3.5.

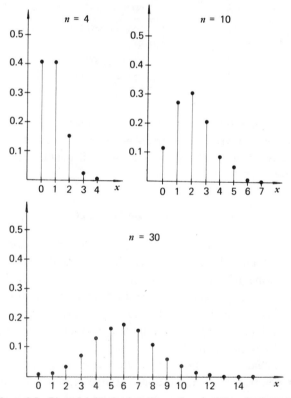

Figure 3.5 Binomial density functions when $\theta = 0.2$ and n is varied.

Computing the binomial probabilities does not encounter difficulties if n is small. However, for large n, either tables of combinatorial coefficients or incomplete Beta functions or an approximation must be employed.

Example. Let us assume that we can say from our experience that the probability θ of a gross error in a certain analytical procedure equals 0.05. If we determine a single analytical result from a series of three determinations, we can notice the probabilities of possible numbers, 0 to 3, of gross errors in such a result:

k	$p(k) = P_3(k)$
0	$(0.95)^3 = 0.8574$
1	$3(0.05)(0.95)^2 = 0.1354$
2	$3(0.05)^2(0.95) = 0.0071$
3	$(0.05)^3 = 0.0001$

One can observe that, for example, the probability of at least one error in one analysis is $1 - 0.8574 = 0.1426$; this means that approximately one-seventh of the analyses will contain at least one gross error.

3.3.2 The Poisson Distribution

The set trials giving rise to the binomial distribution do not involve the factor of time. However, there are random experiments involving time which otherwise fit a binomial model. In those experiments at any instant of time an elementary event occurs or does not. We are interested in the number of these happenings in a given time interval rather than in the number of successes in a prescribed number of trials.

We will name the happenings that occur randomly in time random points. $\xi(t)$ is the random variable representing the number of random points occurring in the interval $\langle 0, t \rangle$. It can take on a countable number of values and its density function can be derived as

$$P\{\xi(t)=k\}=p_t(k)=\frac{(\lambda t)^k}{k!}\exp(-\lambda t) \qquad (3.22)$$

where $\lambda > 0$ is a constant representing the average number of random points per unit time. It is a describing function of the *Poisson distribution*. Its validity requires that these assumptions be satisfied:

1. The probability that k random points occur between t and T depends on k and T but not on t.
2. The occurrence of random points in one interval does not affect the probability of random points in any other nonoverlapping interval.
3. The probability of more than one random point in any interval of length t approaches zero faster than t.

The Poisson density function can also be derived as a limiting case of the binomial distribution by viewing the random phenomenon as a binomial experiment with $n \to \infty$. An observation interval of length t is split into n disjoint time intervals, each of length t/n. Each segment is taken for a trial in the binomial experiment on the basis of assumptions 1 and 2; the occurrence of one or more random points in a segment means a "success" and its probability is θ. Assumption 3 is used in a somewhat different sense. If the number of successes is to equal the number of random points, it must always be possible to split the time axis finely enough to contain no more than one random point in any segment. Assuming that it exists, we can calculate the limit

$$\lim_{n \to \infty} \binom{n}{k}\theta^k(1-\theta)^{n-k}$$

The behavior of θ, as n grows large, is now the focus of the problem. It is desirable to require that the chance of success be approximately equal to a constant, $\lambda > 0$, times the length of the segment:

$$\theta \approx \lambda \frac{t}{n} \qquad \text{as } n \to \infty$$

We can then manipulate the limit above:

$$\lim_{n \to \infty} \binom{n}{k} \left(\frac{\lambda t}{n} \right)^k \left(1 - \frac{\lambda t}{n} \right)^{n-k}$$

$$= \frac{(\lambda t)^k}{k!} \lim_{n \to \infty} \left(1 - \frac{\lambda t}{n} \right)^n \cdot \left(1 - \frac{\lambda t}{n} \right)^{-k} \cdot 1 \cdot \left(1 - \frac{1}{n} \right) \cdots \left(1 - \frac{k-1}{n} \right)$$

$$= \frac{(\lambda t)^k}{k!} \lim_{n \to \infty} \left(1 - \frac{\lambda t}{n} \right)^n$$

$$= \frac{(\lambda t)^k}{k!} \exp(-\lambda t)$$

The last step is a consequence of

$$\lim_{n \to \infty} \left(1 - \frac{\lambda t}{n} \right)^n = \lim_{n \to \infty} \left[\left(1 + \frac{1}{-n/\lambda t} \right)^{-n/\lambda t} \right]^{-\lambda} = e^{-\lambda t}$$

The resulting function is the Poisson density function from (3.22).

The distribution function of this probability distribution can be written

$$F_{\xi(t)}(k) = \sum_i p_t(i) = \sum_i \frac{(\lambda t)^i}{i!} \exp(-\lambda t)$$

where we sum up over all nonnegative integers less or equal to k. Obviously,

$$\sum_{k=0}^{\infty} \frac{(\lambda t)^k}{k!} \exp(-\lambda t) = \exp(-\lambda t) \exp(\lambda t) = 1$$

It remains to prove that $\lambda > 0$ is the expectation of the distribution with unit time and to derive the variance. By definition in (3.17), we have

$$E[\xi(t)] = \sum_{k=0}^{\infty} k p_t(k) = \sum_{k=0}^{\infty} k \frac{(\lambda t)^k}{k!} \exp(-\lambda t)$$

$$= \exp(-\lambda t) \sum_{k=1}^{\infty} \frac{(\lambda t)^k}{(k-1)!}$$

$$= \exp(-\lambda t)\lambda t \sum_{k=1}^{\infty} \frac{(\lambda t)^{k-1}}{(k-1)!}$$

$$= \lambda t \exp(-\lambda t) \exp(\lambda t)$$

$$= \lambda t$$

Putting $t = 1$, we get

$$E[\xi(1)] = \lambda$$

Similarly, the second moment by definition in (3.19) can be adjusted:

$$E[\xi^2(t)] = \sum_{k=0}^{\infty} k^2 p_t(k) = \sum_{k=0}^{\infty} k^2 \frac{(\lambda t)^k}{k!} \exp(-\lambda t)$$

$$= \sum_{k=1}^{\infty} k \frac{(\lambda t)^k}{(k-1)!} \exp(-\lambda t)$$

$$= \lambda t \sum_{m=0}^{\infty} (m+1) \frac{(\lambda t)^m}{m!} \exp(-\lambda t)$$

$$= \lambda t E[\xi(t)] + \lambda t \cdot 1$$

$$= (\lambda t)^2 + \lambda t$$

and for unit time,

$$E[\xi^2(1)] = \lambda^2 + \lambda$$

Thus, by the relation (3.20),

$$V[\xi(t)] = (\lambda t)^2 + \lambda t - (\lambda t)^2 = \lambda t$$
$$V[\xi(1)] = \lambda$$

Hence, λ is not only the average number of random points in a unit interval, but also their variance.

It is important to remark that the discrete random variable in the Poisson distribution assumes a countable number of values, but the probabilities of possible numbers of random points are small. Practically, only probabilities in a few values around λ differ more substantially from zero; those at other points are so small that we can take them for zeros. Tables of Poisson density functions as well as values of Poisson distribution functions for various values of λ can be found in most books on probability and statistics, for example, in [1]. Examples of Poisson density functions are shown in Figure 3.6.

Example.   For λt less than $\ln 2$ (about 0.69), $p_t(0) = \exp(-\lambda t) > 0.5$, so the probability of obtaining no random points during an interval of length t is greater than the chance of obtaining at least one. As λt increases, the probability of obtaining any one value of the Poisson random variable becomes small, although the chance of obtaining a set of values can be large.

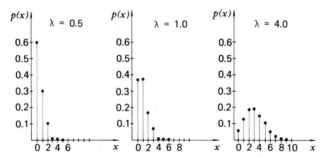

Figure 3.6   Poisson density functions as the parameter is varied.

For instance, we have estimated the average number of pulses per 10-minute intervals from 100 background measurements for low-level β counting as $\lambda t = 10$. From the table of the Poisson density function, we find $p_t(6) = 0.063055$. That is, if the number of pulses that occur in 10-minute intervals were counted for 1.000 such intervals, about 63 of these intervals could be expected to yield six pulses. Although it is possible to observe a very large number of pulses in the given interval, the probability that 25 or more will be observed is 0.000047, or zero to four decimal places. Several events are listed below and their probabilities are computed for $\lambda t = 10$.

$$A_1 = \{\xi(t) > 4\}: \quad \text{at least 5 pulses}$$

$$P(A_1) = \sum_{k=5}^{\infty} p_t(k) = 0.970747$$

$$A_2 = \{\xi(t) \leqslant 8\}: \quad \text{no more than 8 pulses}$$

$$P(A_2) = \sum_{k=0}^{8} p_t(k) = 0.332820$$

3.3.3 The Normal or Gaussian Distribution

The normal, or Gaussian, distribution merits first place among the continuous distributions because it is the most important distribution of probability theory. It is repeatedly encountered in describing the behavior of random variables in nature, industry, and the sciences. It appears also as the distribution of various sample statistics in statistical inference and is the limit distribution of some other probability distributions.

A typical example of the normal distribution is the distribution of random errors in measuring continuous quantities. Similarly, the normal distribution arises when repeating chemical analyses of the same sample or in repeating experiments under the same reaction conditions.

A random variable ξ has a normal, or Gaussian, distribution if its probability density has the form

$$f(x) = \frac{1}{\sqrt{2\pi}\,\sigma} \exp\left[-\frac{1}{2}\left(\frac{x-\mu}{\sigma}\right)^2\right] \qquad x \in \mathcal{R}_1 \qquad (3.23)$$

Thus, it involves the expectation μ and the standard deviation $\sigma > 0$. This frequency function plots into the familiar bell-shaped curve shown in Figure 3.7.

When $\mu = 0$ and $\sigma = 1$, we have the probability density of the standardized normal random variable ζ (see the example in Sec. 3.2.4). The special notation $\phi(z)$ is assigned to its distribution function:

$$\phi(z) = \frac{1}{\sqrt{2\pi}} \int_{-\infty}^{z} \exp\left(-\frac{x^2}{2}\right) dx \qquad (3.24)$$

A table of this function as well as of the probability density

$$\varphi(z) = \frac{1}{\sqrt{2\pi}} e^{-z^2/2} \qquad z \in \mathcal{R}_1 \qquad (3.25)$$

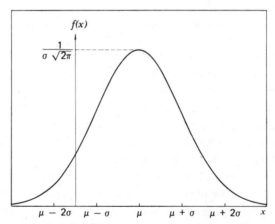

Figure 3.7 Probability density of a normal (Gaussian) distribution.

can be found in any monography on probability and statistics, for example, in [2].

Probabilities defined in terms of the distribution function $F(x)$ of any Gaussian random variable ξ can be computed from $\phi(z)$ as shown in the following derivation.

According to (3.23) and (3.8),

$$F(x) = \frac{1}{\sqrt{2\pi}\,\sigma} \int_{-\infty}^{x} \exp\left[-\frac{1}{2}\left(\frac{y-\mu}{\sigma}\right)^2 \right] dy$$

Applying the change of variable $t = (y - \mu)/\sigma$, $dt = dx/\sigma$, we obtain the integral

$$F(x) = \frac{1}{\sqrt{2\pi}} \int_{-\infty}^{(x-\mu)/\sigma} \exp\left(-\frac{t^2}{2} \right) dt$$

and using (3.24), we have

$$F(x) = \phi\left(\frac{x-\mu}{\sigma}\right) \tag{3.26}$$

The frequency function $\varphi(z)$ has its maximum at the point $z = 0$ and we can easily show that the inflection points have abscissas $z = \pm 1$. Thus, with regard to the transformation

$$\zeta = \frac{\xi - \mu}{\sigma}$$

the frequency function $f(x)$ in (3.23) has its maximum at $x = \mu$ and the inflection points at $x = \mu \pm \sigma$.

Example. The example demonstrates the use of (3.26) in computing probabilities for any normal random variable. The random variable ξ has a normal distribution with $\mu = 100$ and $\sigma^2 = 25$.

Several probabilities will be computed.

$$P\{\xi \leqslant 105\} = F(105) = \phi\left(\frac{105-100}{5}\right)$$
$$= \phi(1) = 0.8413$$
$$P\{\xi > 115\} = 1 - F(115) = 1 - \phi\left(\frac{115-100}{5}\right)$$
$$= 1 - \phi(3) = 1 - 0.99865 = 0.00135$$
$$P\{96 \leqslant \xi \leqslant 108\} = F(108) - F(96) = \phi\left(\frac{108-100}{5}\right) - \phi\left(\frac{96-100}{5}\right)$$
$$= \phi(1,6) - \phi(-0,8)$$

However, the function $\phi(z)$ is tabulated only for nonnegative arguments, and its values at negative points are computed from the symmetry as $\phi(-z) = 1 - \phi(z)$. Hence,

$$P\{96 \leqslant \xi \leqslant 108\} = \phi(1,6) - 1 + \phi(0,8)$$
$$= 0.9452 - 1 + 0.7881 = 0.7333$$

Those values z of the standardized normal random variable will be denoted z_α, for which

$$P\{|\zeta| \leqslant z_\alpha\} = 1 - \alpha \quad \text{and} \quad P\{|\zeta| > z_\alpha\} = \alpha$$

We will call them $100\alpha\%$ critical values (percentage points) of the standardized normal variable. A few of them are computed for several values of α in Table 3.1. Besides some values of α chosen in advance the table involves values of α corresponding to the selected critical values $z_\alpha = 1, 2$, and 3. The probabilities $(1 - \alpha)$ for these values z_α indicate the percentage of the area under the normal frequency function closed, after transformation, between $\mu \pm \sigma, \mu \pm 2\sigma$, and $\mu \pm 3\sigma$, respectively.

One important property of the normal distribution is the fact that any linear combination of independent normal random variables is again distributed normally.

TABLE 3.1 Critical Values of the Standardized Normal Random Variable for Several Values of α

α	$1-\alpha$	z_α
0.50	0.50	0.6745
0.3174	0.6826	1.00
0.10	0.90	1.6448
0.05	0.95	1.9600
0.0454	0.9546	2.00
0.01	0.99	2.5758
0.0026	0.9974	3.00

3.3.4 The Log-Normal Distribution

The log-normal distribution is very useful in statistical work. It has been found, in addition to another use, to fit quite well the distribution of the size of the particles in natural aggregates.

In its simplest form, the log-normal distribution is that of the random variable ξ when the logarithm $\ln \xi$ has a normal distribution with parameters μ and σ^2 We have for its distribution function, by (3.26),

$$F_\xi(x) = P\{\xi \leqslant x\} = P\{\ln\xi \leqslant \ln x\} = \phi\left(\frac{\ln x - \mu}{\sigma}\right)$$

Hence, the probability density is, according to (3.7),

$$f_\xi(x) = F'_\xi(x) = \frac{1}{x\sigma}\varphi\left(\frac{\ln x - \mu}{\sigma}\right)$$

$$= \frac{1}{x\sqrt{2\pi}\,\sigma}\exp\left[-\frac{1}{2\sigma^2}(\ln x - \mu)^2\right] \qquad x > 0$$

$$\text{(3.27)}$$

$$f_\xi(x) = 0 \qquad \text{otherwise}$$

Computing the most important moments we have, by definition in (3.17),

$$E[\xi] = \int_0^\infty \frac{1}{\sqrt{2\pi}\,\sigma} \exp\left[-\frac{1}{2}\left(\frac{\ln x - \mu}{\sigma}\right)^2\right] dx$$

Substituting

$$\frac{\ln x - \mu}{\sigma} = t \qquad \frac{dx}{\sigma} = xt = t\exp(\mu + \sigma t)$$

we get

$$E[\xi] = \frac{1}{\sqrt{2\pi}} \exp\mu \int_{-\infty}^{\infty} \exp\left(\sigma t - \tfrac{1}{2}t^2\right) dt$$

$$= \frac{1}{\sqrt{2\pi}} \exp\left(\mu + \frac{\sigma^2}{2}\right) \int_{-\infty}^{\infty} \exp\left[-\tfrac{1}{2}(t-\sigma)^2\right] dt$$

$$= \exp\left(\mu + \tfrac{1}{2}\sigma^2\right) \tag{3.28}$$

Similarly, we can derive

$$V[\xi] = \exp(2\mu + \sigma^2)(\exp\sigma^2 - 1) \tag{3.29}$$

The log-normal distribution has a mode [the maximum value of $f_\xi(x)$] at the point $x = \exp(\mu - \sigma^2)$.

If not $\ln \xi$, but $\ln(\xi - x_0)$ is normally distributed, the distribution of ξ has the same properties as those described above except that the expected value and the mode are each increased by x_0. Thus, we obtain a "shifted" log-normal distribution.

3.3.5 The Uniform or Rectangular Distribution

The final distribution to be discussed is the uniform or rectangular distribution. It is useful in applications because it implies equally likely outcomes or no prior choice.

ξ is a continuous random variable having a uniform distribution on the interval $\langle a,b \rangle$ if its probability density is given by

$$f(x) = \begin{cases} \dfrac{1}{b-a} & \text{for } a \leqslant x \leqslant b \\ 0 & \text{otherwise} \end{cases} \qquad (3.30)$$

This distribution has the following expected value and variance:

$$E[\xi] = \frac{a+b}{2} \qquad V[\xi] = \frac{(b-a)^2}{12}$$

Sketches of the probability density and the distribution function are shown in Figure 3.8.

Remark. The distribution of the discrete random variable η with the density function

$$p(y) = \begin{cases} \dfrac{1}{n} & \text{for } y = y_i, i = 1, 2,, \ldots, n \\ 0 & \text{otherwise} \end{cases}$$

can be also called uniform.

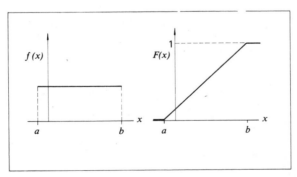

Figure 3.8 Describing functions for uniform (rectangular) distribution.

References

1. E.C. Molina, *Poisson's Exponential Binomial Limit*, D. Van Nostrand Company, New York, 1948.
2. E.S. Pearson and H.O. Hartley, *Biometrika Tables for Statisticians*, Vol. 1, Cambridge University Press, New York, 1958.
3. W. Feller, *An Introduction to Probability Theory and Its Applications*, Vol. 1, J. Wiley & Sons, Inc., New York, 1957.
4. S.S. Wilks, *Mathematical Statistics*, J. Wiley & Sons, Inc., New York, 1963.

INFORMATION THEORY

4.1 HARTLEY'S FORMULA

In our life we receive, store, or transmit information of many kinds, and information is often transformed as it is in telegraphy. If we want to deal mathematically with problems of storing, transferring, and transforming information, we need first to find a quantitative measure of information regardless of the form and the content of the message. It is reasonable to measure the amount of information contained in a message by the number of signs necessary to express its content in the most concise form. The information to be measured must be encoded into the chosen system. It is convenient to choose sequences of zeros and ones for this purpose. The information specifying which of the values 0 and 1 a digit has assumed can be taken as the information unit. We will call this unit a "bit" (an abbreviation of "binary digit").

If we know in advance that an object belongs to a set E, it remains to specify the element of E that is in question, in order to give full information about the object. The amount of information received depends, of course, on the number of elements in E. If E contains exactly $N = 2^n$ elements, we can label them with binary numbers having n digits. Any element will provide n information units. From $N = 2^n$ follows $n = \log_2 N$, from which Hartley derived the idea of defining the necessary information about an element of a set having N elements by $\log_2 N$, even if N is not a power of 2.

At first glance it would seem that if $2^n < N < 2^{n+1}$, then $\log_2 N$ information units will not suffice for the characterization of the elements of E. However, this is not true and it proves that this number is sufficient. Thus, *Hartley's formula*,

$$I(E_N) = \log_2 N \tag{4.1}$$

represents the necessary information about the elements of a set E_N of N elements.

Although the formula (4.1) is a definition of the amount of information, we require that the set function possess certain properties. These postulates are:

1. $I(E_{NM}) = I(E_N) + I(E_M)$ for $N, M = 1, 2, \ldots$.
2. $I(E_N) \leqslant I(E_{N+1})$.
3. $I(E_2) = 1$.

The third requirement is simply the definition of the unit. The meaning of the second postulate is clear: the larger a set, the more information is gained by characterizing its elements. Postulate 1 requires that the information be an additive quantity. (Here we understand that a set E_{NM} can be decomposed into N subsets each of M elements.)

Now it can be shown that these postulates are fulfilled *only* by the function $\log_2 N$. The proof lies beyond the scope of this text, but the function is the starting point for considerations to follow.

4.2 SHANNON'S FORMULA

Let

$$E = \bigcup_{k=1}^{n} E_k$$

be a union of pairwise disjoint finite sets E_k with numbers of elements N_k; E therefore has $N = \sum_{k=1}^{n} N_k$ elements, and we put $p_k = N_k / N$ ($k = 1, 2, \ldots, n$).

The information necessary to characterize an element of E consists of two parts. The first part, I_1, determines the set E_k containing the element in question; the second part, I_2, identifies the element in E_k. Thus, we may write $I = I_1 + I_2$. If we know that an element of E belongs to a set E_k, the remaining amount of information we need equals $\log_2 N_k$. Thus, to characterize such an element we need on the average the amount of information

$$I_2 = \frac{1}{N} \sum_{k=1}^{n} N_k \log_2 N_k = \sum_{k=1}^{n} p_k \log_2 N p_k$$

If we substitute in the equation $I = I_1 + I_2$, we get

$$\log_2 N = I_1 + \sum_{k=1}^{n} p_k \log_2 N p_k$$

$$\log_2 N = I_1 + \log_2 N + \sum_{k=1}^{n} p_k \log p_k$$

since $\sum_{k=1}^{n} p_k = 1$. It follows that

$$I_1 = - \sum_{k=1}^{n} p_k \log_2 p_k = \sum_{k=1}^{n} p_k \log_2 \frac{1}{p_k} \tag{4.2}$$

is the amount of information needed to know to which subset E_k an element of E belongs.

Formula (4.2) was first discovered by Shannon and is called *Shannon's formula*. This formula was also established by Wiener, independently of Shannon.

In particular, if $p_k = 1/n$ for every k ($k = 1, 2, \ldots, n$), Shannon's formula reduces to Hartley's formula [cf. (4.1)].

Shannon's formula was derived implicitly under these assumptions:

1. The selected element from the set E is a value of a random variable.

2. All elements of E are equiprobable; thus, the probability that an element of E belongs to E_k is $p_k = N_k/N$.
3. The partial amounts of information associated with the subsets E_k are "weighted" by the corresponding probabilities; essentially, we consider the expectation of the information in (4.2).

Thus, we now face a more general question: how much information is yielded by the outcome of a random experiment? Will Shannon's formula remain valid in this case?

The problem can be formulated in the following way. Let $s_1, s_2, \ldots, s_n$ be the possible outcomes of a random experiment (points of a finite sample space) and set $p_k = P(s_k)$ for $k = 1, 2, \ldots, n$. We wish to examine the amount of information furnished by a single trial of the experiment.

We must, of course, state in advance some requirements of the information measure. It is reasonable to postulate that:

1. The information obtained depends only on the values of the density function $(p_1, p_2, \ldots, p_n)$; consequently it will be denoted by $I(p_1, p_2, \ldots, p_n)$. We suppose that it is a symmetric function of its arguments.
2. $I(p, 1-p)$ is a continuous function of $p(0 \leqslant p \leqslant 1)$.
3. $I(\frac{1}{2}, \frac{1}{2}) = 1$.

Furthermore, we require that the following relation hold:

4.
$$I(p_1, p_2, \ldots, p_n) = I(p_1 + p_2, p_3, \ldots, p_n)$$
$$+ (p_1 + p_2) I\left(\frac{p_1}{p_1 + p_i}, \frac{p_2}{p_1 + p_2}\right) \quad (4.3)$$

The sense of the last condition is as follows. Suppose that an outcome A of an experiment with probability α can occur in two mutually exclusive ways, A' or A'', the probabilities of which are α' and α'', respectively ($\alpha' + \alpha'' = \alpha$). If we know in which way A actually occurred, the amount of information obtained is associated with the distribution $(\alpha'/\alpha, \alpha''/\alpha)$ taken with the weight α; thus, it equals $\alpha I(\alpha'/\alpha, \alpha''/\alpha)$. Condition 4 also requires the additivity of the information and weighting of information by corresponding probabilities.

Remark. The set of postulates 1 to 4 is a simplified form of those given by Khinchin.

Now it can be proved that the function defined by Shannon's formula in (4.2) is the only function satisfying the given postulates.*

It is useful to add these remarks:

1. If $n=2$, $p_1=1$, and $p_2=0$, it follows from (4.3) that

$$I(1,0) = I(1) + I(1,0)$$

so that $I(1)=0$ (i.e., the occurrence of the sure event does not bring any information).
2. If we obtain some information, the previously existing *uncertainty* will be diminished. Thus, all information obtained means the decrease of uncertainty. Through the outcome of an experiment, not only information but also uncertainty can be measured.
3. The quantity given by Shannon's formula is often called the *entropy* of the probability distribution $(p_1, p_2, \ldots, p_n)$. There is indeed a strong connection between the notion of entropy in thermodynamics and the concept of information (or uncertainty). Boltzmann first emphasized the probabilistic meaning of thermodynamic entropy and, in fact, did pioneering work in information theory. He showed that the entropy of a physical system can be considered to be a measure of the disorder in the system. In some physical systems the numerical values of the disorder also measure the uncertainties concerning the states of

*Since $\lim_{x \to 0+} x \log_2 1/x = 0$, we put $0 \cdot \log_2 1/0 = 0$.

individual particles (as, e.g., in perfect gases). Subsequent development of the concept of entropy extended far beyond physics. There exists a systematic relation between the entropy of macroscopic objects and the measure of available microscopic information regarding them. The objects entering this relation need not be physical objects at all. The idea of entropy found fruitful application in removed fields (mathematics, biology, chemistry, communication theory, philosophy).

4. When we speak of information, we do not have in mind the subjective information of an observer. The information contained in an observation (measurement, determination) is a quantity independent of the perception of an observer (a person, computer, or registering device). Uncertainty is to be interpreted in the same manner.

5. There is still another point which shows that our definition of information is suitable. The unit of information called the "bit" was introduced as being the amount of information contained in a binary symbol. If an experiment is repeated sufficiently often, it can be shown that we do not need, on the average, more than $I(p_1,\ldots,p_n)$ binary symbols for the description of an outcome. Thus, the statement that the outcome of the experiment contains the amount of information $I(p_1,\ldots,p_n)$ has a very specific meaning.

4.3 THE GAIN OF INFORMATION

Now our aim is to compare the introduced information measure for various density functions. For this purpose we will need *Jensen's inequality*: if $g(x)$ is a convex function on the interval (a,b), $x_1, x_2, \ldots, x_n$ are arbitrary real numbers from this interval, and $w_1, w_2, \ldots, w_n$ are positive numbers with $\sum_{k=1}^{n} w_k = 1$, then

$$g\left(\sum_{k=1}^{n} w_k x_k \right) \leqslant \sum_{k=1}^{n} w_k g(x_k) \tag{4.4}$$

The validity of the inequality can be easily seen from the geometrical point of view. Consider the points $(x_k, g(x_k)), k = 1, 2, \ldots, n$, to lie on the convex curve $y = g(x)$ in the plane (x, y) and the masses w_k to be situated on these points. The center of gravity $(\bar{x}, \bar{y})$ will evidently lie above this curve [i.e., $g(\bar{x}) \leqslant \bar{y}$]. Since $\bar{x} = \sum_{k=1}^{n} w_k x_k$ and $\bar{y} = \sum_{k=1}^{n} w_k g(x_k)$, the inequality is proved. If $g(x)$ is not linear on any subinterval, the equality sign in (4.4) cannot occur, except for the case of all points x_k being identical.

If $g(x)$ is concave, the inverse inequality holds because $-g(x)$ is now convex.

If we apply Jensen's inequality to the convex function $y = x \log_2 x (x > 0)$ with $x_k = p_k, w_k = 1/n$ $(k = 1, 2, \ldots, n)$, we get

$$\frac{1}{n} \sum_{k=1}^{n} p_k \log_2 \left(\frac{1}{n} \sum_{k=1}^{n} p_k \right) \leqslant \frac{1}{n} \sum_{k=1}^{n} p_k \log_2 p_k$$

which can be altered to

$$-\frac{1}{n} \log_2 n \leqslant -\frac{1}{n} I(p_1, p_2, \ldots, p_n)$$

$$I(p_1, p_2, \ldots, p_n) \leqslant \log_2 n$$

The equality sign holds only for $p_1 = p_2 = \cdots = p_n = 1/n$. That means that, if an experiment can result in one of n possible outcomes, the uncertainty achieves its maximum when all outcomes are equiprobable.

We introduce now the notion of conditional information. Let ξ and η be two discrete random variables having positive probabilities in the points $x_1, x_2, \ldots, x_m$ and $y_1, y_2, \ldots, y_n$, respectively. We will write

$$
\begin{array}{ll}
P\{\xi = x_i\} = p_i & i = 1, 2, \ldots, m \\
P\{\eta = y_k\} = q_k & k = 1, 2, \ldots, n \\
P\{\xi = x_i, \eta = y_k\} = r_{ik} & \\
P\{\xi = x_i | \eta = y_k\} = p_{i|k} & k = 1, 2, \ldots, n \\
P\{\eta = y_k | \xi = x_i\} = q_{k|i} & i = 1, 2, \ldots, m
\end{array}
$$

According to the definition of conditional probability in (3.1), we have

$$r_{ik} = p_i q_{k|i} = q_k p_{i|k}$$

Furthermore,

$$\sum_{k=1}^{n} r_{ik} = p_i \qquad (i = 1, 2, \ldots, m)$$

and

$$\sum_{i=1}^{m} r_{ik} = q_k \qquad (k = 1, 2, \ldots, m)$$

The *conditional information* $I(\xi|\eta)$ is that contained in ξ given that η has assumed a given value; it is the expectation of the information associated with the distribution $(p_{1|k}, \ldots, p_{m|k})$:

$$I(\xi|\eta) = \sum_{k=1}^{n} q_k I(p_{1|k}, \ldots, p_{m|k})$$

$$= \sum_{k=1}^{m} \sum_{i=1}^{n} r_{ik} \log_2 \frac{q_k}{r_{ik}} \qquad (4.5)$$

On the other hand, if $I[(\xi, \eta)]$ denotes the information associated with the two-dimensional distribution of ξ and η, we can write

$$I[(\xi, \eta)] = \sum_{i=1}^{m} \sum_{k=1}^{n} r_{ik} \log_2 \frac{1}{r_{ik}}$$

Using the preceding relations we see immediately that

$$I[(\xi, \eta)] = I(\eta) + I(\xi|\eta) \qquad (4.6)$$

It follows from the definition that $I(\xi|\eta) = I(\xi)$ when ξ and η are independent; hence, (4.6) obtains the form we used previously:

$$I[(\xi, \eta)] = I(\xi) + I(\eta)$$

The content of the relation in (4.6) is as follows: the information contained in the pair of values of (ξ, η) is the sum of the information contained in the value of η and of the conditional information contained in the value of ξ provided that η has taken on a certain value. This result may be viewed as a generalization of the theorem on the additivity of the information.

Let us return now to the derivation of Shannon's formula. We consider again a set E of N elements and its partition $E_1, E_2, \ldots, E_n$. The subsets E_k have N_k elements, $N = \sum_{k=1}^{n} N_k$ and we put $p_k = N_k / N$. We will label the elements of E from 1 to N [i.e., $E = \{e_1, e_2, \ldots, e_N\}$] and the elements of $E_k (k = 1, 2, \ldots, n)$ from 1 to N_k. An element of E chosen at random may be characterized in two distinct manners (all elements have the same probability $1/N$ of being chosen): (1) by giving its number in the series from 1 to N, we will denote it as ξ; and (2) by giving the set E_k to which the element belongs and its number in the series from 1 to N_k. We will denote the subscript k of E_k as the random variable η and the subscript of the element in the set E_k as ζ. Then we have

$$I(\xi) = I(\eta) + I(\zeta | \eta)$$

where

$$I(\xi) = \log_2 N \qquad I(\eta) = \sum_{k=1}^{n} p_k \log_2 \frac{1}{p_k}$$

and

$$I(\zeta | \eta) = \sum_{k=1}^{n} p_k \log_2 N_k$$

Now we will consider a nonempty subset $E' \subset E$, and $E'_k (k = 1, 2, \ldots, n)$ will be its intersection with E_k. Let N' be the number of elements of E' and N'_k the number of elements of E'_k, and denote $q_k = N'_k / N'$. Then we have $\sum_{k=1}^{n} N'_k = N'$ and therefore $\sum_{k=1}^{n} q_k = 1$. Suppose we know about an element chosen at random that it belongs to E'; we investigate the amount of information that will be furnished hereby about η. The original (a priori) distribution of η was given by the probabilities p_k; after the information that the chosen element belongs to E', η has the a posteriori distribution with probabilities q_k. Thus, we look for the gain of information.

The gained information might seem to equal the difference $I(p_1,\ldots,p_n) - I(q_1,\ldots,q_n)$. However, this difference may be negative while the gain of information may not. The difference is the decrease in uncertainty of η and we look for the gain of information resulting from the knowledge that the element belongs to E'. Let us denote the investigated quantity by $I(q\|p)$; we can determine it by the following reasoning.

The statement $e_\xi \in E'$ contains the information $\log_2(N/N')$. It consists of the information given by the proposition $e_\xi \in E'$ about the value of η and the information given about the value of ζ if η is already known. The second part is easy to calculate; if $\eta = k$, the information obtained equals $\log_2(N_k/N_k')$, and since this information has the probability q_k, the information about the value of ζ is

$$\sum_{k=1}^{n} q_k \log_2 \frac{N_k}{N_k'}$$

Hence,

$$\log_2 \frac{N}{N'} = I(q\|p) + \sum_{k=1}^{n} q_k \log_2 \frac{N_k}{N_k'}$$

Since

$$\sum_{k=1}^{n} q_k = 1 \quad \text{and} \quad \frac{NN_k'}{N'N_k} = \frac{q_k}{p_k}$$

we find that

$$I(q\|p) = \sum_{k=1}^{n} q_k \log_2 \frac{q_k}{p_k} \tag{4.7}$$

Thus, the quantity $I(q\|p)$ depends only on the density functions

$p(x)$ and $q(x)$; it follows from Jensen's inequality that we have always

$$I(q\|p) \geqslant 0 \qquad (4.8)$$

The equality sign holds only if the density functions $p(x)$ and $q(x)$ are identical. The quantity $I(q\|p)$ defined by (4.7) will be called the *gain of information* resulting from the replacement of the a priori distribution $(p_1,\ldots,p_n)$ by the a posteriori distribution $(q_1,\ldots,q_n)$. This is a fundamental notion in information theory.

If $(p_1,\ldots,p_n)$ is any discrete distribution with n nonzero probabilities and $(q_1,\ldots,q_n)=(1/n,\ldots,1/n)$ (a discrete uniform distribution), we have

$$I\left(p\|\frac{1}{n}\right) = \sum_{k=1}^{n} p_k \log_2 np_k = \log_2 n - I(p_1,\ldots,p_n)$$

$$= I\left(\frac{1}{n},\ldots,\frac{1}{n}\right) - I(p_1,\ldots,p_n)$$

Thus, the gain of information obtained by replacing the uniform distribution by the distribution $(p_1,\ldots,p_n)$ equals the decrease of uncertainty. However, in the general case the quantities $I(q\|p)$ and $I(p_1,\ldots,p_n) - I(q_1,\ldots,q_n)$ are different.

The particular case where the initial distribution is uniform plays a role in investigating gains of information in chemical analyses and we will encounter it in Chapter 6.

4.4 EXTENSION OF THE DEFINITION OF INFORMATION

Up to now we have dealt with density functions of finite discrete random variables. They have served us well to introduce basic ideas and concepts of information theory. However, we need to extend information measures to more complex distributions. First, we will be concerned with infinite discrete distributions.

If the random variable ξ assumes a countable number of values x_k with probabilities $P\{\xi = x_k\} = p_k (k = 1, 2, \ldots)$, we define the information contained in the value of ξ by

$$I(\xi) = \sum_{k=1}^{\infty} p_k \log_2 \frac{1}{p_k} \tag{4.9}$$

provided that the series on the right-hand side converges.

If η is a random variable taking on the same values as ξ but having a different distribution with probabilities $P\{\eta = x_k\} = q_k$ ($k = 1, 2, \ldots$), we define the gain of information obtained if the distribution ($p_1, p_2, \ldots$) is replaced by the distribution ($q_1, q_2, \ldots$) as follows:

$$I(q\|p) = \sum_{k=1}^{\infty} q_k \log_2 \frac{q_k}{p_k} \tag{4.10}$$

if the series on the right-hand side converges (which is not always the case).

Now we turn our attention to continuous distributions. We want to extend the definitions of information measures to this case. It cannot be done in a straightforward way, for we would obtain the quantity $I(\xi)$ to be, in general, infinite. More precisely, the amount of information furnished would be infinite if the value of ξ could be known exactly. However, practically, a value of a continuous random variable can be determined only up to a finite number of decimal (or binary) digits.

Thus, we face problems of divergence if we intend to define $I(\xi)$. It will be reasonable to approach a continuous distribution through a discrete one and to investigate the behavior of the information associated with the discrete distribution as the deviation between the two distributions diminishes.

Instead of ξ we can, for instance, consider the random variable

$$\xi_N = \frac{[N\xi]}{N}$$

where $[x]$ denotes the largest integer not exceeding x. We can put

$$p_{N,k} = P\left\{ \xi_N = \frac{k}{N} \right\} = P\left\{ \frac{k}{N} \leqslant \xi < \frac{k+1}{N} \right\}$$

$(k = 0, \pm 1, \pm 2, \ldots; \ N = 1, 2, \ldots)$. If $I(\xi_1)$ is finite, it follows from Jensen's inequality that $I(\xi_N)$ is finite for every N and that

$$I(\xi_N) \leqslant I(\xi_1) + \log_2 N$$

The information measure $I(\xi_N)$ tends to infinity as $N \to \infty$. However, in many cases the limits

$$d(\xi) = \lim_{N \to \infty} \frac{I(\xi_N)}{\log_2 N}$$

$$\lim_{N \to \infty} \left[I(\xi_N) - d(\xi) \log_2 N \right] = I_d(\xi)$$

exist; then the quantity $I_d(\xi)$ represents the d-dimensional information contained in the value of the random variable ξ.

It can be proved that in the important case of absolutely continuous distributions,

$$\lim_{N \to \infty} \frac{I(\xi_N)}{\log_2 N} = 1$$

If

$$\int_{-\infty}^{\infty} f(x) \log_2 \frac{1}{f(x)} \, dx$$

exists, then

$$\lim_{N \to \infty} \left[I(\xi_N) - \log_2 N \right] = I_1(\xi) = \int_{-\infty}^{\infty} f(x) \log_2 \frac{1}{f(x)} \, dx \quad (4.11)$$

is the one-dimensional information measure. It is called also the *entropy* of the continuous random variable ξ having the probability density $f(x)$. The properties of these quantities differ somewhat from those of the information quantities for discrete distributions. For instance, $I_1(\xi)$ can be negative. This is explained by realizing that $I_1(\xi)$ is the limit of a difference between two items of information.

Remark. The subscript 1 in $I_1(\xi)$ will be omitted in applications of this information measure.

It remains for us to mention briefly the notion of the gain of information in the case of continuous distributions. If we have two absolutely continuous probability functions, P and Q, on the real line $\mathfrak{R}_1$ and Q is absolutely continuous with respect to P [i.e., if the probability density $p(x)=0$ at a point x, then $q(x)=0$],

$$I(q\|p) = \int_{-\infty}^{\infty} q(x) \log_2 \frac{q(x)}{p(x)} \, dx \qquad (4.12)$$

is the gain of information obtained if the describing function $p(x)$ is replaced by $q(x)$. This is the quantity we will employ most frequently in our applications. Kullback [3] has considered it as a definition of information for "discriminating in favor of one hypothesis against another hypothesis."

References

1. D. A. Bell, *Information Theory and Its Engineering Applications*, I. Pitman and Sons, London, 1956.
2. L. Brillouin, *Science and Information Theory*, Academic Press, New York, 1962.
3. S. Kullback, *Information Theory and Statistics*, John Wiley & Sons, Inc., New York, 1959.
4. C. E. Shannon, A Mathematical Theory of Communication, *Bell Syst. Tec. J.*, **27** (1948).
5. D. A. S. Fraser, *Ann. Math. Stat.*, **36**, 890 (1965).
6. A. Rényi, *Probability Theory*, Akad. Kiadó, Budapest, 1970.
7. A. I. Khinchin, *Mathematical Foundations of Information Theory*, Dover, New York, 1957.

SOME METHODS OF STATISTICAL INFERENCE

5.1 THE RANDOM SAMPLE

The basic objective of mathematical statistics is to present inference conclusions emerging from the knowledge obtained from the set of observed values of one or several random variables. Such a set can in most cases be extended. The set of results of all possible repetitions of a random experiment will be called a *population*. For our needs in analytical chemistry problems, infinite and hypothetical populations are of greatest importance. We can name all possible values of the concentration of a compound that we could obtain under unchanged conditions, or the results of all possible analyses that we could carry out by the same analytical method on the same sample.

We have to distinguish between a sample space and a population. For instance, in repeating infinitely many times a random experiment with only two possible outcomes denoted 0 and 1, the sample space $S = \{0, 1\}$ is finite while the population is infinite (made of zeros and ones, i.e., records of the outcomes).

Examination of large populations in their entirety is impossible. Therefore, we try to investigate them on the basis of their parts, that is, using samples. If we determine values of a random variable in a series of n independent repetitions of a random experiment with which this random variable is associated, we get a *random sample* of values $x_1, x_2, \ldots, x_n$, with x_i being the value of the random variable in the ith repetition. We can state the following definition.

Definition. A random sample of size n is an n-dimensional random variable $(\xi_1, \xi_2, \ldots, \xi_n)$ whose components are independent and have the same distribution function $F(x)$. Thus, the probability distribution of the random sample is described according to the extension of (3.15a) of Section 3.2.3 by the joint distribution function

$$H(x_1, x_2, \ldots, x_n) = \prod_{i=1}^{n} F_i(x_i) = [F(x)]^n$$

It is evident that this definition expresses random sampling from an infinite population in which the probability distribution remains unchanged during the sampling. There also exists the theory of random samples from finite populations, which we can, of course, easily omit.

Remark.   In the following discussion we will no longer distinguish the notation of a random sample (as of a random variable) from the notation of its value; both will be denoted by $(x_1, x_2, \ldots, x_n)$.

In this chapter we will first be concerned with distributions of some sample statistics that play a leading role in both the theory of estimates and statistical tests. Three additional probability distributions, important in statistical inference, are discussed in the Appendix.

5.2   PROBABILITY DISTRIBUTIONS OF SAMPLE STATISTICS

In sampling theory we investigate probability distributions of one or several functions of n random variables forming a random sample. Any sample statistic (e.g., mean, sample variance) is such a function $g(x_1, x_2, \ldots, x_n)$, which is itself a random variable. Its probability distribution will be called a *sample distribution* and this distribution is fully determined by the probability distribution in the population from which the sample is taken. Thus, we will speak, for instance, of the sample distribution of the means; and so on.

5.2.1   The Sample Distribution of the Mean

Suppose that $(x_1, x_2, \ldots, x_n)$ is a random sample from a population whose distribution has expected value μ and variance σ^2. Then according to the properties of the expectation in Section 3.2.4, we have

$$E[\bar{x}] = \frac{1}{n}(E[x_1] + \cdots + E[x_n])$$

But since x_i $(i = 1, 2, \ldots, n)$ are random variables having identical distribution functions, we have

$$E[x_1] = \cdots = E[x_n] = \mu$$

Thus, we get

$$E[\bar{x}] = \frac{1}{n}n\mu = \mu \qquad (5.1)$$

Without attempting a discussion of the basic concepts of statistical estimation, we remark that (5.1) asserts that $\bar{x}$ is an *unbiased estimator* for μ.

Now consider the variance of $\bar{x}$. According to the properties of the variance stated in Section 3.2.4, we have

$$V[\bar{x}] = \frac{1}{n^2}(V[x_1] + \cdots + V[x_n])$$

However,

$$V[x_i] = \sigma^2 \qquad (i = 1, 2, \ldots, n)$$

and thus

$$V[\bar{x}] = \frac{1}{n^2} n\sigma^2 = \frac{\sigma^2}{n} \qquad (5.2)$$

We have shown that the variance of the mean is n times less than the variance in the population.

If the probability distribution of the population is normal, the distribution of $\bar{x}$ is also normal, because $\bar{x}$ is a linear combination of random variables x_i (see Section 3.3.3). If the sample is taken from a population with another probability distribution, the remarkable *central limit theorem* can be applied. Under fairly general conditions, this theorem implies that the sum of any sequence of stochastically independent random variables can be approximated by a normal distribution. Thus, $\bar{x}$ is also asymptotically normally distributed, and the approximation by the normal distribution fits well generally for small values of n ($n \geqslant 4$).

5.2.2   The Sample Distribution of the Sample Variance

Now consider the sample variance, which is defined as

$$s^2 = \frac{1}{n-1} \sum_{i=1}^{n} (x_i - \bar{x})^2 \qquad (5.3)$$

We can write

$$s^2 = \frac{1}{n-1} \sum_{i=1}^{n} \left[(x_i - \mu) - \frac{1}{n} \sum_{j=1}^{n} (x_j - \mu) \right]^2$$

$$= \frac{1}{n} \sum_{i=1}^{n} (x_i - \mu)^2 - \frac{1}{n(n-1)} \sum_{i \neq j} (x_i - \mu)(x_j - \mu)$$

Taking the expectations and being aware of

$$E\left[(x_i - \mu)(x_j - \mu) \right] = 0 \qquad (i \neq j)$$

we have

$$E[s^2] = \frac{1}{n} \sum_{i=1}^{n} E[(x_i - \mu)^2] = \frac{1}{n} n\sigma^2 = \sigma^2 \qquad (5.4)$$

We remark that the reason for using $(n-1)$ as the divisor in (5.3) rather than n is to make $E[s^2]$ exactly equal to σ^2, that is, to make s^2 an *unbiased estimator* for σ^2.

Carrying out similar operations with the expectation, we find that

$$V[s^2] = \frac{1}{n} \left(\mu_4 - \frac{n-3}{n-1} \sigma^4 \right) \qquad (5.5)$$

where μ_4 is the fourth central moment of the population distribution.

Very little is known about the sampling distribution of s^2 except for the case of sampling from normally distributed populations. However, we are primarily interested in this case with regard to applications in analytical chemistry.

If the samples of size n are taken from a normal population with a variance σ^2, it can be determined that the random variable $(n-1)s^2/\sigma^2$ has the χ^2-distribution with $(n-1)$ degrees of freedom (see the Appendix). Moreover, this random variable, as well as s^2 itself, is stochastically independent of the mean, $\bar{x}$.

5.2.3 The Student Ratio

We have seen in Section 5.2.1 that sample means in samples from a normal population are normally distributed with parameters $(\mu, \sigma^2/n)$. Thus,

$$\frac{\bar{x} - \mu}{\sigma} \sqrt{n}$$

is a standardized normal random variable. Besides, as we know,

$$\frac{(n-1)s^2}{\sigma^2}$$

is a χ^2-random variable with $(n-1)$ degrees of freedom, and these two random variables are independent. We can then form (according to the Appendix) a random variable having the Student t-distribution. Thus,

$$t = \frac{\dfrac{\bar{x}-\mu}{\sigma}\sqrt{n}}{\sqrt{\dfrac{(n-1)s^2}{\sigma^2(n-1)}}} = \frac{\bar{x}-\mu}{s}\sqrt{n} \tag{5.6}$$

is a Student random variable with $(n-1)$ degrees of freedom. It is known as the *Student ratio*; it is important that it does not contain the parameter σ.

5.3   CONFIDENCE INTERVALS

One of the objectives of statistical inference is to estimate unknown parameters of the probability distribution in the population from which a random sample originates. For analytical chemistry the most valuable from the practical point of view are interval estimates (i.e., we try to find such an interval that will include the unknown value of the parameter with a probability given in advance). Such intervals are called *confidence intervals*. The probability α $(0 < \alpha < 1)$ is called the *confidence level*, and it represents the risk that the interval will not include the true value of the parameter. Confidence intervals can be obtained by various methods. Generally, the shortest ones are preferred.

It will be sufficient for our needs to find confidence intervals for the expected value in a normal distribution. The problem will be split into two parts.

1. We assume that the variance σ^2 is known. The ratio

$$\frac{\bar{x}-\mu}{\sigma}\sqrt{n}$$

is a standardized normal random variable, the values of which are tabulated [because the means $\bar{x}$ are normally distributed with parameters $(\mu, \sigma^2/n)$]. According to the definition of critical values in Section 3.3.3, z_α is such a value that, for a given α,

$$P\left\{-z_\alpha \leqslant \frac{\bar{x}-\mu}{\sigma}\sqrt{n} \leqslant z_\alpha\right\} = 1-\alpha$$

This value can be found in the table of the normal distribution function $\phi(z)$. The inequality can be manipulated into

$$P\left\{\bar{x} - \frac{\sigma z_\alpha}{\sqrt{n}} \leqslant \mu \leqslant \bar{x} + \frac{\sigma z_\alpha}{\sqrt{n}}\right\} = 1-\alpha \qquad (5.7)$$

and we have thus obtained a $100(1-\alpha)\%$ confidence interval for the unknown expected value μ. The interval is symmetrical around $\bar{x}$, and it is the shortest one of all intervals with the given confidence level.

Sometimes one-sided confidence intervals are more desirable. Since $P\{\zeta > z_{2\alpha}\} = P\{\zeta < -z_{2\alpha}\} = \alpha$ as the consequence of the symmetry of the standardized normal distribution, then

$$P\left\{\frac{\bar{x}-\mu}{\sigma}\sqrt{n} \leqslant z_{2\alpha}\right\} = P\left\{\frac{\bar{x}-\mu}{\sigma}\sqrt{n} \geqslant -z_{2\alpha}\right\} = 1-\alpha$$

It follows that

$$P\left\{\bar{x} - \frac{\sigma z_{2\alpha}}{\sqrt{n}} \leqslant \mu\right\} = P\left\{\bar{x} + \frac{\sigma z_{2\alpha}}{\sqrt{n}} \geqslant \mu\right\} = 1-\alpha \qquad (5.8)$$

and we have thus derived one-sided $100(1-\alpha)\%$ confidence intervals for μ.

2. We assume that the variance σ^2 is unknown. Then obviously the Student ratio from Section 5.2.2 can be used, for it does not contain the parameter σ. We have, for a given α,

$$P\left\{-t_\alpha(n-1) \leqslant \frac{\bar{x}-\mu}{s}\sqrt{n} \leqslant t_\alpha(n-1)\right\} = 1-\alpha$$

where $t_\alpha(n-1)$ is the critical value of the Student distribution with $(n-1)$ degrees of freedom and can be taken from the table of these values (see reference in the Appendix). The inequality can be rearranged so that

$$P\left\{\bar{x}-\frac{st_\alpha(n-1)}{\sqrt{n}} \leqslant \mu \leqslant \bar{x}+\frac{st_\alpha(n-1)}{\sqrt{n}}\right\}=1-\alpha \quad (5.9)$$

and we obtain the confidence interval in question.

Example. A series of six determinations of the concentration of alkali hydroxide solution yielded the values: 68.5%, 69.2%, 68.6%, 68.2%, 68.8%, and 68.9%. The results are assumed to be a sample from a large number of possible concentrations, in which the frequency distribution would fit the frequency curve of a normal distribution. We want to compute the confidence interval for the real value of the concentration. From the six determinations we have calculated $\bar{x}=68.70$ and $s=0.35$. Since the critical value taken from the table is $t_{0.05}(5)=2.57$, the 95% confidence interval for the unknown μ equals

$$68.70-\frac{2.57\times0.35}{\sqrt{6}} \leqslant \mu \leqslant 68.70+\frac{2.57\times0.35}{\sqrt{6}}$$
$$68.33 \leqslant \mu \leqslant 69.07$$

Thus, we can state that the interval $\langle 68.33\%; 69.07\%\rangle$ includes the true value of the concentration of the given solution, with a 5% risk that this statement is false.

Remark. For a given method, the size of a confidence interval depends on the probability α and on the sample size n. As far as the confidence level α is concerned, we have to choose between a more general statement, which has a high probability of being correct, and a more specific statement, which has a smaller probability of being correct. For a fixed value of α, we obtain shorter confidence intervals when n is large than when n is small.

We do not present confidence intervals for other parameters and other distributions in the populations because we prefer to discuss some of them in the next section from the point of view of the theory of statistical decision making.

5.4 TESTING STATISTICAL HYPOTHESES

One important question is whether a sample of observations can be taken for a random one originating from a population with a certain probability distribution. We want to know whether the deviations of the empirical distribution from the hypothetical one can be assigned to random factors or whether they are significant and indicate a real difference between the two distributions. In another case the probability distribution in the population is considered to be known (e.g., from a physical law or from experience) except for one or several parameters. It is then useful to formulate a hypothesis of the values of these parameters and verify it by the means of an available random sample. The manner of verifying a hypothesis is called a *significance test*.

This section is devoted to the basic principles of the theory of testing parametric statistical hypotheses and to those applications of them which occur in connection with evaluating analytical results and methods. We encounter these problems whenever we are to determine whether a new process gives a greater yield than an old one, whether a given analytical method gives correct results when employed to analyze a known content, whether a given shipment of raw material contains less than a certain amount of impurity, and so on.

If the distribution function in the population depends on one parameter θ, we will denote it as $F(x|\theta)$, and a random sample of size n will have the distribution function $[F(x|\theta)]^n$. If a hypothesis is of the form $\theta = \theta_0$, it is called *simple*; otherwise, it is *composite*. A test procedure simply amounts to a rule whereby each possible set of values $(x_1, x_2, \ldots, x_n)$ is associated with a decision to accept or reject the hypothesis H_0 being tested. We can visualize

$(x_1, x_2, \ldots, x_n)$ as a point in the Euclidean n-dimensional space and then the points for which H_0 is rejected are regarded as forming the *critical region* of the test.

It is obvious that errors of two kinds may be committed in applying a statistical test:

1. Errors of type I: rejecting H_0 when it is true.
2. Errors of type II: accepting H_0 when some other hypothesis is true.

The probability of the error of type I is

$$\alpha = P\{(x_1, x_2, \ldots, x_n) \in w | H_0\} \qquad (5.10)$$

where w is the critical region. Alpha is often made equal to 0.05 or 0.01, although other values can be used. To facilitate the construction of such tests, tables have been prepared for some standard cases. The probability α is called the *significance level* of the test.

The probability of a type II error depends on the hypothesis H that is really true. It is equal to

$$\beta = 1 - P\{(x_1, x_2, \ldots, x_n) \in w | H\} \qquad (5.11)$$

and is called the *operating characteristic* of the test with respect to H. The complement

$$P\{(x_1, x_2, \ldots, x_n) \in w | H\} \qquad (5.12)$$

is called the *power* of the test with respect to H.

The decision procedure for a test of H_0 is shown in Table 5.1. Note in this table that if H_0 is true and is accepted, the decision is correct. If it is false but is accepted, the decision produces a type II error. If H_0 is true but is rejected, a type I error arises.

By decreasing the extent of w, we reduce the significance level α. This also results in an increase in the probability of a type II error. There thus exists a conflict involving reduction of the probabilities of the first and second types of error. To resolve this

TABLE 5.1   Decisions and Errors

Hypothesis H_0	Decision	
	Accept H_0	Reject H_0
Is true	Correct	α
Is false	β	Correct

conflict we try to control the significance level and, subject to this control, to minimize β (i.e., to maximize the power of the test with respect to an alternative hypothesis H). This will be illustrated in the application of the test procedure to a hypothesis regarding the size of the expected value in a normal distribution.

Remark.   If a hypothesis is accepted, the experimenter may still not be willing to say that it is true. What he may say is that on the prescribed significance level he has not enough evidence from the obtained sample to reject the hypothesis provided that it is false.

5.4.1   Tests of Significance for the Mean Value of a Normal Distribution

As in our first case we will investigate the test of significance on the mean value when the underlying distribution of the data is normal and when a random sample of size n is available. We have to distinguish again whether the variance σ^2 is known or not.

1. The variance σ^2 is assumed to be known. We have recognized that

$$\frac{\bar{x} - \mu}{\sigma} \sqrt{n}$$

is a standardized normal variable. We choose α and state the hypothesis H_0: $\mu = \mu_0$. We know (see Section 3.3.3) that the probability that

$$|z_0| = \left| \frac{\bar{x} - \mu_0}{\sigma} \sqrt{n} \right| > z_\alpha$$

is equal α provided that H_0 is true. Since α is small, we reject the hypothesis H_0 if this happens. Otherwise, we do not reject it. Thus, the union of intervals $(-\infty, -z_\alpha)$ and (z_α, ∞) is the critical region.

Suppose now that we have accepted H_0 but that another hypothesis H_1: $\mu = \mu_1$ is true (we have thus committed a type II error). Then, of course,

$$z_1 = \frac{\bar{x} - \mu_1}{\sigma} \sqrt{n}$$

is a standardized normal variable and we can manipulate it as

$$z_1 = \frac{\bar{x} - \mu_0}{\sigma} \sqrt{n} - \frac{\mu_1 - \mu_0}{\sigma} \sqrt{n}$$
$$= z_0 - d\sqrt{n}$$

where

$$d = \frac{\mu_1 - \mu_0}{\sigma}$$

Thus, we have for the probability of the error of type II,

$$\beta = P\left\{ -z_\alpha \leqslant \frac{\bar{x} - \mu_0}{\sigma} \sqrt{n} \leqslant z_\alpha \right\}$$
$$= P\left\{ -z_\alpha \leqslant z_1 + d\sqrt{n} \leqslant z_\alpha \right\}$$
$$= P\left\{ -z_\alpha - d\sqrt{n} \leqslant z_1 \leqslant z_\alpha - d\sqrt{n} \right\} \qquad (5.13)$$

Since z_1 is now a standardized normal variable, β can be found by the use of tables of normal distribution. Obviously, as n increases, both interval limits in (5.13) increase or decrease (according to whether d is positive or negative) and β tends to zero.

The test we have just investigated is a two-sided test. Sometimes, however, it is desirable that H_0 not be accepted only if μ

is in fact greater than μ_0. Then the critical region with the significance level α will be given by

$$\frac{\bar{x}-\mu_0}{\sigma}\sqrt{n} > z_{2\alpha}$$

The operating characteristic will change, for if $H_1: \mu = \mu_1 > \mu_0$ is true, we have

$$\beta = P\left\{\frac{\bar{x}-\mu_0}{\sigma}\sqrt{n} \leqslant z_{2\alpha}\right\} = P\left\{z_1 \leqslant z_{2\alpha} - d\sqrt{n}\right\} \quad (5.14)$$

The operating-characteristic curve for this test is illustrated in Figure 5.1 for a few values of n. It is clear that β is small if μ_1 is substantially greater than μ_0. If $\mu_1 < \mu_0$, the hypothesis H_0 will be accepted even more frequently than in the case $\mu = \mu_0$.

Another way of looking at the errors of both types for this test is illustrated in Figure 5.2. If H_1 is true, $\bar{x}$ has the frequency

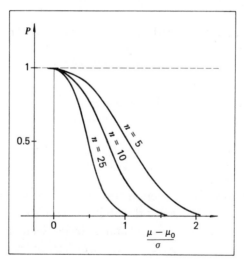

Figure 5.1 Operating characteristic curves for the test of $H_0: \mu = \mu_0$ against $H_1: \mu = \mu_1 > \mu_0$, $\alpha = 0.01$, σ^2 known.

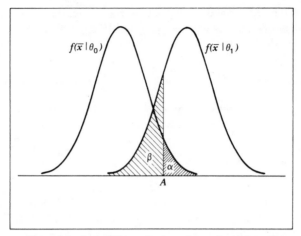

Figure 5.2 Probabilities of errors of type I and type II.

function $f(\bar{x}|\theta_1)$ and

$$\beta = \int_{-\infty}^{A} f(\bar{x}|\theta_1)\,d\bar{x}$$

which is the area marked by thin crosshatching. The probability

$$\alpha = \int_{A}^{\infty} f(\bar{x}|\theta_0)\,d\bar{x}$$

equals the densely crosshatched area.

2. The variance σ^2 is not known. Then the statistic

$$\frac{\bar{x}-\mu_0}{s}\sqrt{n}$$

might be used as a criterion. Its distribution when H_0 is true ($\mu = \mu_0$) is the Student t-distribution with $(n-1)$ degrees of freedom (it is the Student ratio from Section 5.2.3). The hypothesis will be rejected on the significance level α whenever

$$\left|\frac{\bar{x}-\mu_0}{s}\sqrt{n}\right| > t_\alpha(n-1) \tag{5.15}$$

if we test it against an alternative hypothesis H_1: $\mu = \mu_1$. This is again a two-sided test. In a similar way we would get one-sided significance tests.

However, the power of these tests using the t-statistics cannot be evaluated without knowing σ, so we cannot, in fact, construct a test satisfying the conditions required above.

Remark. Tests on the mean values in normal distributions allow us to decide whether a method in chemical analysis, verified on a sample with a standard content, is "biased" or not. The standard value is taken as μ_0.

5.4.2   Comparison of Two Mean Values in Normal Distributions

A frequent problem in analytical chemistry is to compare two mean values on the basis of independent samples from normally distributed populations with these parameters.

1. Both variances σ_1^2 and σ_2^2 are assumed to be known. Here, we calculate $\bar{x}_1$ from a sample of size n_1 and $\bar{x}_2$ from a sample of size n_2. Then the difference $\bar{x}_1 - \bar{x}_2$ is, according to Sections 3.3.3 and 5.1, normally distributed with $\mu = \mu_1 - \mu_2$ and $\sigma^2 = \sigma_1^2/n_1 + \sigma_2^2/n_2$. If we set the hypothesis H_0: $\mu_1 = \mu_2 \Rightarrow \mu_1 - \mu_2 = 0$, the statistic

$$\frac{\bar{x}_1 - \bar{x}_2}{\sqrt{\sigma_1^2/n_1 + \sigma_2^2/n_2}} \tag{5.16}$$

is a standardized normal variable. The two-sided test is carried out in the usual manner.

2. Both variances σ_1^2 and σ_2^2 are unknown. Here, only a statistic with an approximate probability distribution can be applied to serve as a test criterion. If we denote

$$s_{\bar{x}_1}^2 = \frac{s_1^2}{n_1} \qquad s_{\bar{x}_2}^2 = \frac{s_2^2}{n_2}$$

then the statistic

$$\frac{\bar{x}_1 - \bar{x}_2}{\sqrt{s_{\bar{x}_1}^2 + s_{\bar{x}_2}^2}} \tag{5.17}$$

has approximately a Student t-distribution if the hypothesis H_0: $\mu_1 = \mu_2$ is true. The number of degrees of freedom m is taken from the equation

$$\frac{1}{m} = \frac{c^2}{n_1 - 1} + \frac{(1 - c)^2}{n_2 - 1} \quad \text{with} \quad c = \frac{s_{\bar{x}_1}^2}{s_{\bar{x}_1}^2 + s_{\bar{x}_2}^2}$$

and rounded. The calculated value is compared with $t_\alpha(m)$ and the hypothesis H_0 is rejected if

$$\frac{|\bar{x}_1 - \bar{x}_2|}{\sqrt{s_{\bar{x}_1}^2 + s_{\bar{x}_2}^2}} > t_\alpha(m)$$

5.4.3   Comparison of Two Variances in Normal Distributions

If s_1^2 and s_2^2 are sample variances in two independent random samples taken from normally distributed populations, then according to Section 5.2.2, the statistics

$$\frac{n_1 - 1}{\sigma_1^2} s_1^2 \quad \text{and} \quad \frac{n_2 - 1}{\sigma_2^2} s_2^2$$

are independent random variables having χ^2-distributions with $(n_1 - 1)$ and $(n_2 - 1)$ degrees of freedom, respectively. Then, according to Appendix A.3, the ratio

$$F = \frac{s_1^2/\sigma_1^2}{s_2^2/\sigma_2^2} = \frac{\sigma_2^2}{\sigma_1^2} \frac{s_1^2}{s_2^2} \tag{5.18}$$

has the F-distribution with a pair of degrees of freedom (n_1-1, n_2-1).

To compare the variances σ_1^2 and σ_2^2, it is convenient to state the simple hypothesis H_0: $\sigma_1^2=\sigma_2^2$. If it is true, the ratio

$$\frac{s_1^2}{s_2^2}$$

has the F-distribution; thus, we compare it with the critical value $F_\alpha(n_1-1,n_2-1)$. If it exceeds this critical value, we reject H_0.

We are usually interested in testing H_0 against H_1: $\sigma_1^2>\sigma_2^2$. For instance, we would prefer an analytical method the precision of which is expressed by σ_1^2, without proving that it is less precise (i.e., that $\sigma_1^2>\sigma_2^2$) than another method.

If $\sigma_1^2<\sigma_2^2$, the random variable in (5.18) has the F-distribution and we have

$$P\left\{\frac{s_1^2}{s_2^2}>F_\alpha(n_1-1,n_2-1)\right\}=P\left\{F>\frac{\sigma_2^2}{\sigma_1^2}F_\alpha(n_1-1,n_2-1)\right\}$$
$$<P\{F>F_\alpha(n_1-1,n_2-1)\}=\alpha$$

Thus, we can conclude that, if in fact $\sigma_1^2<\sigma_2^2$, the probability of a type I error is even less than α. Therefore we can state a composite hypothesis H: $\sigma_1^2\leq\sigma_2^2$ and use the critical region $F>F_\alpha(n_1-1, n_2-1)$. The probability of committing a type I error (i.e., to conclude that $\sigma_1^2>\sigma_2^2$) is at most equal to α.

5.5 LINEAR REGRESSION ANALYSIS

We have been concerned so far with one-dimensional random samples and their statistics in relation to their parent populations. Now we will draw our attention to couples of observations (x_i,y_i), for which a special model is assumed. We consider the y_i's to be values of independent *random variables*, whereas the x_i's are referred to as *fixed values*.

Such cases arise, for example, in comparing results obtained by a new analytical method on samples the composition of which is well known, or in observing the dependence of the quality of a product on carefully controlled temperature. If the results of an experiment depend on time, we can usually measure time with sufficient precision and thus consider its measured values as fixed.

Suppose that $y_1, y_2, \ldots, y_n$ are n independent random variables having variances all equal to σ^2 but with expectations given by the *regression function*

$$E[y_i | x_i] = \alpha + \beta x_i \qquad (5.19)$$

where x_i are known (fixed) values but α and β are unknown parameters, called *regression coefficients*, to be estimated. The geometrical interpretation of (5.19) is that the expected values of n normal distributions lie on a straight line.

The estimators of the regression coefficients α and β can be found by the least-squares method. It is known from statistical theory that these estimates appear to be minimum variance estimates. Thus, we minimize the quadratic form

$$Q = \sum_{i=1}^{n} (y_i - \alpha - \beta x_i)^2$$

By solving the necessary conditions for the minimum, we obtain for the estimators a and b,

$$b = \frac{\sum_{i=1}^{n} (x_i - \bar{x})(y_i - \bar{y})}{\sum_{i=1}^{n} (x_i - \bar{x})^2} \qquad (5.20)$$

$$a = \bar{y} - b\bar{x} \qquad (5.21)$$

Obviously, the estimated (empirical) regression line passes through the point $(\bar{x}, \bar{y})$.

The estimators a and b were derived without any assumptions concerning the probability distribution of the random variables y_i.

However, if further inferences are to be drawn, we have to add the assumption of normality of the y_i's. The model can then be shown as in Figure 5.3. Since $\tilde{y}_i = a + bx_i$ is the estimate of $E[y_i|x_i]$, the statistic

$$s_{y|x}^2 = \frac{\sum\limits_{i=1}^{n} (y_i - \tilde{y}_i)^2}{n-2} \qquad (5.22)$$

is taken as the estimate of σ^2 unless it is known. The denominator $(n-2)$ is explained by two degrees of freedom which we lost by estimating α and β from the sample. Under the assumption of normality, $(n-2)s_{y|x}^2 / \sigma^2$ has a χ^2-distribution with $(n-2)$ degrees of freedom (cf. Section 5.2.2).

It can be easily shown that the estimator b is also a normal random variable with

$$E[b] = \beta \qquad V[b] = \frac{\sigma^2}{\sum\limits_{i=1}^{n} (x_i - \bar{x})^2}$$

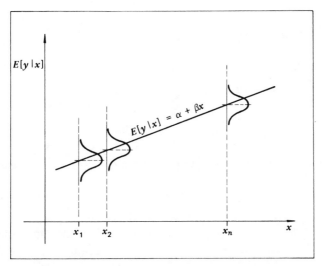

Figure 5.3 Graphic representation of the regression model.

Thus, b is an unbiased estimate of β. Since, in addition, b and $s_{y|x}^2$ are independent, the statistic

$$t = \frac{[(b-\beta)/\sigma]\sqrt{\sum_i (x_i - \bar{x})^2}}{\sqrt{s_{y|x}^2/\sigma^2}} = \frac{b-\beta}{s_b} \qquad (5.23)$$

has, according to the Appendix, the Student t-distribution with $(n-2)$ degrees of freedom. It can be employed in the usual manner to test the hypothesis H_0: $\beta = \beta_0$. The $100(1-\alpha)\%$ confidence interval for β is

$$\langle b - t_\alpha(n-2)s_b; b + t_\alpha(n-2)s_b \rangle$$

Similarly, the estimator a is an unbiased estimate of α having the variance

$$V[a] = \sigma^2 \left[\frac{1}{n} + \frac{\bar{x}^2}{\sum_{i=1}^{n} (x_i - \bar{x})^2} \right]$$

and the $100(1-\alpha)\%$ confidence interval for α is bounded by

$$a \pm t_\alpha(n-2)s_{y|x} \sqrt{\frac{1}{n} + \frac{\bar{x}^2}{\sum_i (x_i - \bar{x})^2}}$$

The empirical regression value $\tilde{y}_0 = a + bx_0$ serves for prediction of the expected value of a random variable y_0 associated with a fixed value x_0.

Remark. The model in (5.19) can be extended in such a way that values of a vector $(x_1, x_2, \ldots, x_k)$ condition the expected value of the random variable. This is the case of the multiregression function.

5.6 SIMPLEX OPTIMIZATION

This section relates to the optimization of analytical procedures to be discussed in Chapter 6. Its objective is to outline the theoretical background of optimization methods. These methods belong, of course, to multivariate statistical analysis and should rather follow a section on experimental designs. Yet, very little is required as a prerequisite to understanding the outline in this section.

Factorial designs, in which all factors are varied simultaneously, are in common use in analytical chemistry. However, these designs have some deficiencies, as has been pointed out, for example, in [3]. They are very sensitive to the choice of tested levels when a factor is continuous; they can explore either a small region comprehensively or a large region superficially; and they may explore regions that are of no interest because they are far from optimum.

If the purpose of an experimental design is to achieve an optimum, then the sequential single-factor approach, which is common in analytical chemistry, might be better than the classic factorial design. This approach requires all factors but one to be held constant while the remaining factor is varied until an optimum response is obtained. The optimized factor is then held constant and a different factor is searched until a new optimum response is obtained; and so on. Several cycles of varying the factors are usually necessary to precisely find the optimum. However, this approach can be deceptive if the response surface contains a ridge [4].

The *sequential simplex method* [5] forces the pattern to move from a given subregion in the direction of steepest ascent as estimated by response measurements at the points in the pattern. The optimum thus obtained will, in general, be a local optimum; if the search is repeated starting from a different initial region, we can verify if the optimum is also global.

The rules of the modified simplex method are shown in Figure 5.4. These simplex designs are conceptually different from the simplex methods used in linear programming.

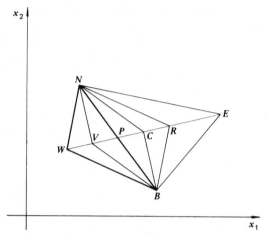

Figure 5.4 Simplex operation.

A simplex is a geometric figure defined by $(n+1)$ points, which is equal to one more than is the number of dimensions of the factor space (n). In the initial simplex BNW shown in Figure 5.4, the response at the vertex B was found to be the best, W the worst, and N the next to worst. P is the center of the face remaining when the worst vertex is removed. The method now consists of reflection, expansion, and contraction rules.

Reflection is accomplished by extending the line segment WP beyond P to generate the new vertex, R:

$$R = P + (P - W)$$

Then three possibilities exist for the measured response at R:

1. The response at R is more desirable than the response at B. The segment PR is then expanded (usually by 2). If the response at point E is again better than the response at B, it is retained as the new simplex BNE; if it is worse, the expansion is considered to have failed and BNR is taken as the new simplex, which is used to restart the algorithm.
2. The response at R is between those of B and N. Then the process is restarted with the new simplex BNR.

3. The response at R is less desirable than the response at N. This indicates that a wrong direction has been chosen and the simplex has to be contracted:

   a. If the response is between those of N and W, the segment PR is contracted (usually by 0.5) and a new point C is obtained. The process is restarted with the new simplex BNC.

   b. If the response at R is even worse than the previous worst vertex W, a new vertex V is chosen between W and P and we restart with the new simplex BNV.

If the contractions result in poorer responses, the size of the simplex can be diminished further.

If a vertex lies outside the boundaries of at least one factor, a very undesirable response is assigned to that vertex and the simplex is forced back inside the boundaries.

The simplex is halted when the step size becomes less than a predetermined value or when the differences in response cannot be distinguished from possible errors.

The multivaried response surface approach, applying a modified simplex method to a colorimetric method, has been examined in [6].

References

1. S. S. Wilks, *Mathematical Statistics*, John Wiley & Sons, Inc., New York, 1963.
2. N. L. Johnson and F. C. Leone, *Statistics and Experimental Design in Engineering and the Physical Sciences*, Vol. 1, John Wiley & Sons, Inc., New York, 1968.
3. M. Friedman and L. J. Savage, in *Techniques of Statistical Analysis*, McGraw-Hill Book Company, New York, 1947.
4. G. E. P. Box, *Biometrics*, **10**, 16 (1954).
5. W. Spendley, G. P. Hext, and F. R. Himsworth, *Technometrics*, **4**, 441 (1962).
6. S. L. Morgan and S. N. Deming, *Anal. Chem.*, **46**, 1170 (1974).

EVALUATION AND OPTIMIZATION OF ANALYTICAL RESULTS AND METHODS

6.1 EMPLOYING INFORMATION THEORY IN ANALYTICAL CHEMISTRY

If we define analysis as a process of obtaining information about the chemical composition of matter and analytical method as the system in which this information is produced, it then appears reasonable to view both analytical results and analytical methods, methodology, and individual procedures from the standpoint of their quantitatively expressible information properties [1, 2].

We primarily use the *information content* or the *information amount* as characteristics for evaluating information properties of analytical results. It is possible to judge analytical methods according to their information properties either absolutely (i.e., to evaluate individual methods as themselves) or relatively (i.e., to consider the suitability of their use for a certain concrete analytical task). In *absolute evaluation* of an analytical method, we consider either the information amount that the method is able to provide in a time unit (i.e., we will use information performance as the basis of its evaluation) or we take into account the cost of obtaining a unit of information (i.e., we will use the specific prices of the information for characteristics of analytical methods [3]). As a suitable quantity for *relative judging* of an analytical method (i.e., for evaluating the expediency of its use for a given task), the information profitability proves competent [4].

The assessment of analytical results and the absolute evaluation of analytical methods has importance for comparison and potential classification. The relations for determining the information content, performance, or specific price of the information can also be used as mathematical models of the process of obtaining information about the chemical composition of the analyzed sample. By discussing them we can investigate conditions under which the results have a high information content or the analytical method exhibits its maximum information performance or works with the lowest specific information price. Such a discussion of the influence of conditions upon information properties of the results is surely very useful, yet for practical needs, relative assessment (i.e., the assessment of the suitability of the use of a particular analytical method for a given task) is far more appropriate. Also, the optimization of an analytical procedure must always be carried out with respect to its use for a concrete purpose.

The *optimization process* can basically be split into two steps: (1) the choice of the most suitable method for the given task from the set of all analytical methods applicable to its solution; and (2) the choice of conditions under which the selected analytical method provides results that have for the specific goal the most suitable properties. The properties of a method are characterized by an objective function, so in the optimization procedure we are seeking the conditions under which the objective function exhibits its extreme (i.e., its maximum or minimum) value. In the optimizing process we can use such quantities for an objective function as the information profitability. It assesses an analytical method or procedure relatively, with regard to its use for a concrete goal, and characterizes the quantitative part of the information obtained, enabling us to also take into account its qualitative part (e.g., the relevance). It has its own importance, particularly in the choice and optimization of an analytical method that is to be used in routine analytical quality control, for clinical examinations, for environmental investigations, and so on. Sometimes a characteristic such as information profitability can also be useful in determining the optimum equipment for a laboratory which is concerned with performing analyses on a certain type of material or specialized analyses in a certain direction; in determining the expediency of automatic processing of analytical data; in determining suitable apparatus; and so on.

In the following sections we will be concerned with relations based on the divergence information measure [1], since it complies best with the theoretical bases of analysis as a process of obtaining information, and because it conforms to practical use, especially by enabling us to derive expressions for the information content in various sorts of analyses.

Remark. The term "information content" will be used in this chapter rather than "information gain," which was used in Chapter 4, and will be denoted by $I(p,p_0)$ rather than $I(p\|p_0)$.

6.2   INFORMATION CONTENT OF A QUALITATIVE PROOF

The results of qualitative analysis are by nature alternative information which tells us whether a certain a priori expected component is or is not present in the sample in an amount corresponding to the detection limit of the method being used. The information content of a simple equality

$$I(p,p_0) = \ln \frac{k_0}{k} \qquad (6.1)$$

where k_0 and k are numbers of possible but unidentified components expected prior to the proof (k_0) or possible after it (k), follows from the use of the divergence measure in (4.12) except for the base of the logarithm. For computational reasons, we will prefer natural logarithms and express information contents in natural units ("nits"). At the same time we assume that the probability of occurrence is equal for all the components. It is obvious that k_0 and k are always integers and that $k_0 \geqslant k$, $k > 0$.

The use of (6.1) will be demonstrated by a simple example. For instance, in a sample only one of the following 20 cations can be present: Ag^+, Pb^{2+}, Hg^{2+}, Cd^{2+}, Bi^{3+}, As^{3+}, Sb^{3+}, Sn^{2+}, Al^{3+}, Zn^{2+}, Fe^{2+}, Mn^{2+}, Ca^{2+}, Sr^{2+}, Ba^{2+}, Mg^{2+}, Na^+, K^+, Li^+, or NH_4^+. If the precipitation proof by dilute HCl has a positive result, then $k = 2$ (Ag^+ or Pb^{2+}) and the information content of this proof is found as follows: $I(p,p_0) = \ln(20/2) = 2.303$ natural information units. However, if the result is negative, then $k = 20 - 2 = 18$, and any of the cations, except Ag^+ and Pb^{2+}, can be contained in the sample. Here, the information content of the proof is only $I(p,p_0) = \ln(20/18) = 0.105$ unit. The information content of the qualitative proof is thus also conditioned by its result and by the selectivity of this proof.

If various combinations of the components are possible, k and k_0 have to be computed by the use of combinatorial techniques. The number of combinations of m components from the total number of M components without repetition ($M \geqslant m$) is given by

$$k_m(M) = \binom{M}{m} = \frac{M!}{m!(M-m)!} \qquad (6.2a)$$

The total number of all the combinations is then

$$k = \sum_{m=0}^{M} k_m(M) = \sum_{m=0}^{M} \frac{M!}{m!(M-m)!} \qquad (6.2b)$$

If one to six of the cations Ag^+, Pb^{2+}, Zn^{2+}, Al^{3+}, Na^+, and K^+ can be in the sample, then $k_1(6) = 6$, $k_2(6) = 15$, $k_3(6) = 20$, $k_4(6) = 15$, $k_5(6) = 6$, and $k_6(6) = 1$, so $k = 63$. In 15 of all possible combinations, neither Ag^+ nor Pb^{2+} is contained, and the reaction with dilute HCl is therefore negative. The information content of the negative reaction with dilute HCl is thus $I(p,p_0) =$

$\ln(63/15) = 1.435$ units; if the reaction is positive (i.e., if either Ag^+ or Pb^{2+} or both are present in the sample), then $I(p,p_0) = \ln(63/48) = 0.272$ unit.

If we perform qualitative analysis by the use of an instrumental method, the information content of such a proof is determined according to (6.1) as well, but the values k_0 and k have to be determined with respect to the analytical method being used. For instance, in emission spectrometry we can prove 32 commonly occurring elements in the range 2000 to 5000 Å. If we assume 18 elements in the sample, of which 2 cannot be determined by emission spectrometry, then $k_0 = 18$ and $k = 2$ and the information content of the spectrometric analysis of the given sample is always $I(p,p_0) = \ln(18/2) = 2.197$ natural information units, regardless of how many elements we proved in the sample. In qualitative analysis, when all assumed components are equally probable in occurrence, finding a component, as opposed to eliminating the presence of a component, each has the same information content. Qualitative analysis using an instrumental analytical method is usually performed simultaneously with quantitative analysis (see Section 6.8). The problems of the information content of qualitative proofs are treated by H. Malissa [5] and K. Danzer [6]; the treatises of the latter are directed to procedures of qualitative analysis rarely used today.

6.3   INFORMATION CONTENT OF A QUANTITATIVE DETERMINATION

The results of a quantitative determination are most frequently normally distributed, and before the analysis we do not know anything more about the content of the determined component than that $X_i \in \langle x_1, x_2 \rangle$. In an extreme case, this can be $x_1 = 0\%$ and $x_2 = 100\%$. If the result of the analysis confirms the a priori assumptions of the content of the component to be determined [i.e., if we find that $(x_1 + 3\sigma) \leqslant \mu \leqslant (x_2 - 3\sigma)$ and if $(x_2 - x_1) \geqslant 6\sigma$, where σ^2 is the variance of the normally distributed results], the information content of the quantitative determination expressed by the use of the divergence measure is given by

$$I(p,p_0)=\ln\frac{x_2-x_1}{\sigma\sqrt{2\pi e}} \tag{6.3}$$

Thus, we obtain the unit information content when $(x_2-x_1)=\sigma e\sqrt{2\pi e}$ [i.e., approximately when $(x_2-x_1)\approx11.234\sigma$]. In practice, we usually determine the result of a quantitative analysis as the average of n_p parallel determinations; in addition, we do not know the true value of the parameter σ but only its estimate,

$$s=\sqrt{\frac{1}{n_s-1}\sum_i(x_i-\bar{x})^2}$$

We then have to express the information content of the result of the quantitative analysis as

$$I(p,p_0)=\ln\frac{(x_2-x_1)\sqrt{n_p}}{2st(\nu)} \tag{6.4}$$

where $t(\nu)$ is the critical value of the Student distribution for the number of degrees of freedom $\nu=n_s-1$ and on the significance level $\alpha=0.038794$. This significance level is chosen in such a way that it holds for $\nu\to\infty$: $2t(\infty)=\sqrt{2\pi e}=4.132731$ and $t(\infty)=2.066366$. The critical values of the Student t-distribution for this α are tabulated in Table 6.1. In practical calculations, especially when using a pocket or table calculator, direct computing the critical values $t(\nu)$ is more advantageous than using the table. Values $t(\nu)$ for $\alpha=0.038794$ can be determined with sufficient precision by the means of the approximation $t(\nu)=\sum_{i=0}^5 a_i(\frac{1}{\nu})^i$. The values of the coefficients $a_i(i=0,1,\ldots,5)$ are as follows*: $a_0=2.0664$, $a_1=2.7250$, $a_2=3.4223$, $a_3=4.0707$, $a_4=0.2123$, $a_5=3.8933$.

---

*This approximation for the computation of critical values of the Student t-distribution is due to I. Horsák.

TABLE 6.1 Critical Values of the Student t-Distribution for $\alpha = 0.038794$

ν	$t(\nu)$
1	16.3899
2	4.9282
3	3.5244
4	3.0296
5	2.7824
6	2.6351
7	2.5377
8	2.4686
9	2.4171
10	2.3773
11	2.3455
12	2.3196
13	2.2981
14	2.2800
15	2.2645
16	2.2510
17	2.2393
18	2.2290
19	2.2199
20	2.2117
21	2.2043
22	2.1977
23	2.1916
24	2.1861
25	2.1811
30	2.1611
35	2.1471
40	2.1367
45	2.1286
50	2.1222
60	2.1127
80	2.1009
100	2.0939
150	2.0847
200	2.0801
∞	2.066366

The computation is easy provided that the Horner's scheme is employed:

$$t(\nu) = a_0 + \frac{1}{\nu}\left(a_1 + \frac{1}{\nu}\left(a_2 + \frac{1}{\nu}\left(a_3 + \frac{1}{\nu}\left(a_4 + \frac{1}{\nu}a_5\right)\right)\right)\right).$$

The values of $I(p,p_0)$ for various s, n_s, and n_p are given in Table 6.2.

The computation of the information content of the result of a quantitative analysis will be shown by a simple example. Prior to an analysis we assume that the content of the determined component lies between $x_1 = 10\%$ and $x_2 = 50\%$; the result $\mu = 14.68\%$ confirmed the assumption. If we know that $\sigma = 0.05\%$, the information content of one determination $I(p,p_0) = \ln 40/0.05\sqrt{2\pi e} = 5.266$ natural information units and the content of the mean of three determinations $I(p,p_0) = \ln 40\sqrt{3}/0.05\sqrt{2\pi e} = 5.815$ units. If we did not know the value of the parameter σ and estimated it by means of s from a series of $n_s = 12$ parallel determinations carried out independently of the analysis of this example, and if

TABLE 6.2 Information Content of Quantitative Analysis Results:

$$I(p,p_0) = \ln \frac{(x_2 - x_1)\sqrt{n_p}}{2st(\nu)}$$

				s				
		0.100				0.001		
				n_s				
n_p	n_p	10	24	∞	n_p	10	24	∞
2	5.303	5.710	5.783	5.835	9.909	10.315	10.388	10.440
3	5.658	5.913	5.985	6.038	10.262	10.518	10.590	10.643
4	5.886	6.057	6.129	6.182	10.491	10.662	10.734	10.787
5	6.052	6.168	6.241	6.294	10.657	10.773	10.846	10.899
6	6.181	6.259	6.332	6.385	10.786	10.864	10.937	10.990
7	6.285	6.336	6.409	6.462	10.890	10.942	11.014	11.067
8	6.373	6.403	6.476	6.529	10.978	11.008	11.081	11.134
9	6.449	6.462	6.535	6.588	11.054	11.067	11.140	11.193
10	6.515	6.515	6.587	6.640	11.120	11.120	11.132	12.091

we found $s = 0.063\%$ with $t(11) = 2.3455$, the information content of one determination would be

$$I(p,p_0) = \ln \frac{40}{0.063 \times 2 \times 2.3455} = 4.908 \text{ units}$$

and the content of the mean of three parallel determinations would be

$$I(p,p_0) = \frac{40\sqrt{3}}{0.063 \times 2 \times 2.3455} = 5.457 \text{ units}$$

If an estimate s calculated from independent determinations were not available, then in carrying out three parallel determinations we would have $n_p = n_s = 3$, and the information content for the case when $s = 0.063\%$ and $v = n_s - 1 = 2$, $t(2) = 4.9282$, would be

$$I(p,p_0) = \ln \frac{40\sqrt{3}}{0.063 \times 2 \times 4.9282} = 4.715 \text{ natural information units}$$

This example illustrates the influence of the number of parallel determinations performed.

Equations (6.3) and (6.4) are often cited in the analytical literature [2,3,7–12]. Here we will note, in addition, the influence of the conditions under which we perform the analysis on the information content of its result:

1. The less concrete are the preliminary assumptions of the true content X_i of the component determined, the greater is the information content of the result. Yet it must always hold $6\sigma \leqslant (x_2 - x_1) \leqslant 100\%$, since for intervals $\langle x_1, x_2 \rangle$ that are narrower than 6σ equations (6.3) and (6.4) are not applicable.
2. The information content given by (6.3) or (6.4) in the case $(x_1 + 3\sigma) \leqslant \mu \leqslant (x_2 - 3\sigma)$ does not depend on the value μ found for the component. This is self evident, of course, since the information about the observed value having obtained its assumed magnitude has a zero content. However, the information content can depend on the content X_i of the component

at that time if the size of σ varies with μ, which is a rather frequent case.

3. The information content depends on the precision and accuracy of the results and on the number of determinations n_p and n_s; this dependence will be treated in Section 6.4.

4. The information content of a quantitative analysis depends on the way in which the calibration was performed and the extent to which the precision and accuracy depend on it. Details will be given in Section 6.5.

The case of the information content of a quantitative analysis whose results do not confirm the original assumption (i.e. the case $\mu \notin \langle x_1 + 3\sigma_1 x_2 - 3\sigma \rangle$) has been solved in [8]. Here we need to evaluate the divergence integral with respect to chosen integral boundaries and to replace the a posteriori normal distribution by a truncated one. The information content is then given by

$$I(p,p_0) = \ln \frac{x_2 - x_1}{[\phi(z_2) - \phi(z_1)]\sqrt{2\pi e}\ \sigma} + \frac{1}{2} \frac{z_2\varphi(z_2) - z_1\varphi(z_1)}{\phi(z_2) - \phi(z_1)}$$

(6.5)

where ϕ is the distribution function from (3.24), φ is the frequency function from (3.25), and $z_i = (x_i - \mu)/\sigma$ for $i = 1, 2$. In this case the information content depends on μ when $z_2 \to 0+$ or $z_1 \to 0-$. Thus, for example, if $\sigma = 0.02$ and $x_2 - x_1 = 100\sigma = 2$, we get for $z_2 = |z_1| = 50$ (i.e. when $\mu = (x_1 + x_2)/2$) the information content $I(p,p_0) = 3.186$ nat. units; for $z_2 = 3$ we have $I(p,p_0) = 3.194$ nat. units, for $z_2 = 1$, $I(p,p_0) = 3.503$ nat. units, and for the case $\mu = x_2 \Rightarrow z_2 = 0$, $I(p,p_0) = 3.879$ nat. units. However, if μ is outside the interval $\langle x_1, x_2 \rangle$ it is necessary to consider repeating the determination with another standard addition or with another sensitivity or even by using another analytical method, more appropriate for the true value X. The case when the results of the analysis do not confirm the original expectation is important in analytical quality control (see Section 6.10 and [13]).

6.4   INFLUENCE ON INFORMATION CONTENT OF NUMBER OF OBSERVATIONS, PRECISION, AND ACCURACY

The information content of the results of a quantitative analysis expressed in terms of the divergence measure by (6.4) depends on their precision, which is characterized by the value of the sample standard deviation s, on the number of parallel determinations n_p and the critical value of the Student distribution $t(\nu)$ $(\nu = n_s - 1)$ on an a priori given significance level of $\alpha = 0.038794$. Here n_s is the number of determinations, from which the standard deviation was estimated. This estimate is given basically by the applied analytical method and is somewhat influenced by the level of the work in the laboratory; it is more reliable the greater the number of determinations n_s.

As far as the accuracy of the results is concerned, we can consider either the bias δ (see Section 2.6) or its estimate $d = |\bar{x} - X|$. Then two cases can arise: (1) the error d is not statistically significant and can be accounted for by the presence of random errors even if in reality $X_i = \mu_i$; and (2) the error d is statistically significant and the applied analytical method provides results subject to a systematic error. The philosophy of the significance tests and the tests for the difference $|X_i - \bar{x}_i|$ were introduced in Chapter 5. If $d > 0$ does not differ significantly from zero, we express the information content of the results of a quantitative analysis by (6.3) or (6.4). Yet, if δ must be taken for positive (i.e., a systematic error is present), the information content of the results is given by the use of the divergence measure as

$$I(p, p_0) = \ln \frac{x_2 - x_1}{\sigma \sqrt{2\pi e}} - \frac{1}{2} \left(\frac{\delta}{\sigma} \right)^2 \qquad (6.6)$$

This expression becomes (6.3) for $\delta = 0$; details can be found in [68]. The dependence of $I(p, p_0)$ on both precision and accuracy is illustrated in Figure 6.1, where it is apparent that the information

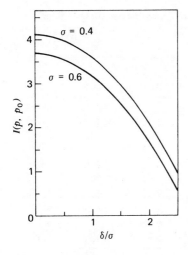

Figure 6.1 Information content of biased results.

content of the results of quantitative analyses is greater the better their precision and accuracy. The values of $I(p,p_0)$ according to (6.6) are tabulated in Table 6.3 for various σ and δ. The information content in the case $\delta = 0$ is expressed in units identical with "bit" or "nit" units as they are used in communication theory. If $\delta \neq 0$, $I(p,p_0)$ given by (6.6) depends on both σ and δ, and the information units can be defined by the means of the *unit isoinform* [i.e., a curve connecting points of $I(p,p_0) = 1$ for corresponding pairs of values σ and δ; see Figure 6.2]; the corresponding pairs of σ and δ for a unit isoinform are given in Table 6.4.

The information content of the results of quantitative analyses depends not only on s (if σ is unknown) but also on the number n_s of results from which this estimate was computed, since the critical value $t(\nu)$ of the Student distribution in (6.4) is a function of $\nu = n_s - 1$. Moreover, the information content given by (6.4) depends on the number n_p of parallel determinations. Regarding the mutual relation of n_p and n_s we can consider two cases: (1) $n_p = n_s$ (i.e., we compute s from the parallel determinations obtained by the applied analytical method of unknown σ); and (2) $n_p < n_s$ (i.e., we carry out fewer parallel determinations than the number of

TABLE 6.3   Information Content of Quantitative Analysis Results:

$$I(p,p_0)=\ln\frac{x_2-x_1}{\sigma\sqrt{2\pi e}}-\frac{1}{2}(\frac{\delta}{\sigma})^2 \qquad \textbf{for} \quad x_2-x_1=100$$

	σ					
δ/σ	0.500	0.100	0.050	0.010	0.005	0.001
0.00	3.879	5.489	6.182	7.791	8.485	10.094
0.05	3.878	5.488	6.181	7.790	8.484	10.093
0.10	3.874	5.484	6.177	7.786	8.480	10.089
0.20	3.859	5.469	6.162	7.771	8.465	10.074
0.40	3.799	5.409	6.102	7.711	8.405	10.014
0.60	3.699	5.309	6.002	7.611	8.305	9.914
0.80	3.559	5.169	5.862	7.471	8.165	9.774
1.00	3.379	4.989	5.682	7.291	7.985	9.594
1.50	2.754	4.364	5.057	6.666	7.360	8.969
2.00	1.879	3.489	4.182	5.791	6.485	8.094

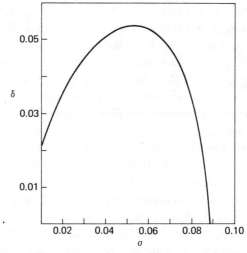

Figure 6.2   Isoinform curve for the information content of biased results.

TABLE 6.4   Isoinform for

$$I(p,p_0) = \ln \frac{1}{\sigma\sqrt{2\pi e}} - \frac{1}{2}\left(\frac{\delta}{\sigma}\right)^2 = 1$$

σ	δ/σ
0.005	0.0120
0.010	0.0209
0.020	0.0346
0.030	0.0442
0.040	0.0506
0.050	0.0537
0.054	0.0540
0.060	0.0533
0.070	0.0485
0.080	0.0370
0.089	0.0000

those from which the standard deviation was estimated in another series). The second case is common in routine applications of analytical methods in quality control, clinical diagnostics, and so on. For the case $\delta = 0$, the dependence of $I(p,p_0)$ on n_p, shown in Table 6.5, is illustrated in Figure 6.3 for $n_p = n_s$ and $n_p \ll n_s$. The great difference in information content in these two cases, particularly for a small value of n_p, is striking. This difference can be logically interpreted such that information about the precision of the results of the applied analytical method obtained from a series of n_s independent determinations and condensed in the value of s is also employed in further determinations obtained by this method, thereby increasing the information content of the results. In practice, also worthwhile is the finding that an initial considerable increase in information content with increasing values of n_p and n_s has later diminished and for high values of n_p and n_s has remained almost constant. This can be explained by the fact that large numbers n_p and n_s contribute to the increase of information redundance rather than to its content.

TABLE 6.5 **Information Content of Quantitative Analysis Results:**

$$I(p,p_0) = \ln \frac{(x_2 - x_1)\sqrt{n_p}}{\sigma \sqrt{2\pi e}} \qquad \text{for} \quad x_2 - x_1 = 100$$

n_p			σ			
	0.500	0.100	0.050	0.010	0.005	0.001
1	3.879	5.489	6.182	7.791	8.485	10.094
2	4.225	5.835	6.528	8.137	8.831	10.440
3	4.428	6.038	6.731	8.340	9.034	10.643
4	4.572	6.182	6.875	8.484	9.178	10.787
5	4.684	6.294	6.987	8.596	9.290	10.899
6	4.775	6.385	7.078	8.687	9.381	10.990
7	4.852	6.462	7.155	8.764	9.458	11.067
8	4.919	6.529	7.222	8.831	9.525	11.134
9	4.978	6.588	7.281	8.890	9.484	11.193
10	5.030	6.640	7.333	8.942	9.636	12.091

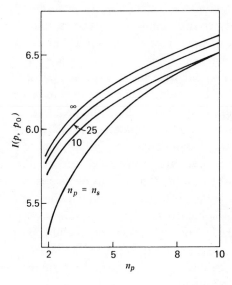

Figure 6.3 Information content of quantitative analysis results with $n_s = n_p$, 10, 25, and ∞.

6.5   INFLUENCE ON INFORMATION CONTENT OF CALIBRATION OF ANALYTICAL METHODS

The information content of the results of quantitative analyses depends, as was shown in the preceding section, primarily on the precision and accuracy of the results. The precision of the results is characterized by the value of σ or by its estimate s and the accuracy by the value of the mean error δ or by the value of its statistically significant estimate d. The values s and d are inherent in certain analytical methods but are influenced to a great extent by the level of the work of the laboratory, the precision and accuracy of the results of methods that require empirical determination of the function g_i from (2.9) or from (2.11a) or (2.11b) of Section 2.5, and the manner in which the calibration is carried out.

Since the purpose of calibration is to obtain accurate results, we will assume, with certain simplifications, a perfect calibration such that the results are not subject to a systematic error (i.e., that $\delta = 0$ or the value d is statistically insignificant). Thus, we take into account the influence of calibration upon the information content of the result of a quantitative determination for these cases: (a) the calibration is carried out by the use of a standard addition; and (b) the calibration is carried out by constructing a calibration straight line. The first case involves calibration by the use of a single standard, and the second, the use of a number of standards.

The method of standard addition is carried out such that we perform a determination with an analyzed sample and obtain a signal of intensity y_1, and we repeat the determination after we have added the standard, obtaining a signal of intensity $y_2 > y_1$. Then the difference $y_s = y_2 - y_1$ corresponds to the standard addition x_s and the result of the analysis

$$x_1 = \frac{y_1}{y_s} x_s = x_s \frac{y_1}{y_2 - y_1}. \tag{6.7a}$$

The standard deviation of the result obtained in this way is of the size

$$s = \frac{s_y \, x_s}{(y_2 - y_1)^2} \sqrt{y_1^2 + y_2^2} \qquad (6.7b)$$

where s_y is the standard deviation of the determination of the signal intensity. The smaller the value of s, the less is s_y, which also depends on the ratio $y_2 : y_1$. The question of the optimum ratio was discussed earlier [15]; approximately, it is optimal for y_1 between $0.4y_2$ and $0.5y_2$. The loss of information content arising by breaking the conditions of the optimum ratio $y_2 : y_1$ in the standard-addition method is treated in [16]. An advantage of the method is its simplicity; it removes almost completely the possibility of a rise in the systematic error when a good blank correction was made; a disadvantage is that it is applicable only if the regression relationship between X_i and $E[\eta_i]$ is perfectly linear and for $X_i = 0$ is $E[\eta_i] = 0$. The results of the determinations obtained by the use of the standard-addition method are less precise, since the value of the square root in (6.7b) can be, usually when the ratio $y_2 : y_1$ is undesirable, quite large. The problem of finding the precision of the results when the method of a standard addition is used was treated by Larsen et al. [14].

Calibration carried out by use of an analytical calibration curve constructed of N points, each of which corresponds to another concentration of the component to be determined, enables us to determine the result x_i of the analysis putting

$$x_i = x_i^{(c)} \qquad (6.7c)$$

where $x_i^{(c)}$ is the amount of the component to be determined, obtained from the calibration straight line. If we repeat the determination n_p times, the standard deviation of the result is computed by (with $\bar{y}(n_p) = \left(\sum\limits_{j=1}^{n_p} y_{m+j} \right) / n_p$)

$$s = \frac{s_y}{b_c} \sqrt{\frac{1}{n_p} + \frac{1}{N} + \frac{(\bar{y}(n_p) - \bar{y})^2}{b_c^2 \sum\limits_i (x_i - \bar{x})^2}} \qquad (6.7d)$$

where b_c, the slope of the calibration curve, is a measure of the sensitivity defined in Section 2.3. Obviously, the information content of a determination carried out by use of a calibration curve increases the greater are the number n_p of parallel determinations and the number N of points of which the calibration curve is constructed, the more precise is the determination of the intensity of the signal (i.e., the smaller s_y is), and the more sensitive is the determination (i.e., the greater the slope b_c). The number of points N must be at least three; after all, it is quite common in analytical practice to construct a calibration straight line of five or more points. The advantage of the use of calibration curve is frequently good precision of the results and possibility of being used for nonlinear dependence. A disadvantage is that we have to maintain exactly the conditions under which the determinations were carried out, and there is a certain risk that a systematic error will occur if the standards and the analyzed sample have different compositions and the results of the applied analytical method show a considerable matrix effect. Other details can be found in [16].

Next we note the influence of the calibration upon the information content of the results of a method subject to a nonzero mean error δ, assuming that the calibration completely removes the mean error but that the precision of the determination somewhat deteriorates at the same time, so that the initial value of the standard deviation σ_1 of the determination performed before the calibration changes by introducing the calibration (e.g., by a correction) into $\sigma_2 > \sigma_1$. If we compare (6.3) by the use of σ_2 with

(6.6) by the use of σ_1, it is obvious that the information content will be the same in both cases when

$$\delta = \sqrt{2\sigma_1^2 \ln \frac{\sigma_2}{\sigma_1}} = \delta_0 \qquad (6.8a)$$

so that it is reasonable to carry out the calibration if $\delta \geqslant \delta_0$. Since for the case that $\sigma_1 \ll 1$ and $\sigma_1 \approx \sigma_2$, δ_0 is very small, which means that in practical work whenever we find a statistically significant mean error d, it is reasonable to carry out the calibration. Comparison of the information content of the results obtained by the use of various methods of calibration can be useful wherever we must decide on the most suitable method. Although the choice of the most appropriate procedure is the subject of Section 6.12, we give here an example concerning the choice of the most suitable method of calibration.

If we carry out a determination in which the precision of the determination of the signal intensity is expressed by the value $s_y = 0.012$, we get for the information content, by use of the standard-addition method at $y_1 = 0.45$ and $y_2 = 1.00$, $I(p,p_0) = 4.43$ information units. In using the calibration curve for $b_c = 1$, $N = 8$, and $n_p = 2$ when the value of $d = 0.01$, the information content $I(p,p_0) = 5.33$ natural information units and is thus greater than that in use of the standard-addition method. As far as the mean error (which has been found to be $d = 0.01$ in a series of 12 determinations) is concerned, we can verify by the use of the t-test that it is statistically insignificant on the level $\alpha = 0.05$ and that it cannot be taken for a systematic error. However, if we found an error $d = 0.025$ in 12 determinations, we would have $t = 9.126 > t(12) = 2.201$; hence, the error would be significant on the level $\alpha = 0.05$ and the information content would be $I(p,p_0) = 3.66$ natural information units. Then it would be desirable to adopt the method of standard addition, in spite of the results being less precise; on the other hand, they would not be subject to a systematic error.

Sometimes it is possible to remove or at least reduce the systematic error for $\mu_i > X_i$ by subtracting a blank correction, which is, however, accompanied by certain deterioration of the precision of the results. The result of the analysis is then determined from the difference in the intensity of the signal y_i obtained in the determination itself and the intensity of the signal y_0 for the blank correction. Since the determination itself and the blank correction run independently, $\sigma = \sqrt{\sigma_i^2 + \sigma_0^2}$, where σ_i^2 is the variance of the proper determination and σ_0^2 the variance in the blank correction. In the rather frequent case $\sigma_i = \sigma_0$, we have $\sigma = \sigma_0 \sqrt{2}$, and it is then expedient to subtract the value of the blank correction if

$$\ln \frac{x_2 - x_1}{\sigma_0 \sqrt{2\pi e}} - \frac{1}{2} \left(\frac{\delta}{\sigma_0} \right)^2 \leqslant \ln \frac{x_2 - x_1}{2\sigma_0 \sqrt{\pi e}}$$

that is, in the case when the mean error that can be removed or reduced by subtracting the blank correction

$$\delta^{(0)} \geqslant \sigma_0 \sqrt{\ln 2} \approx 0.83 \sigma_0 \tag{6.8b}$$

Any experimentally determinable value of the blank correction is, of course, always greater than the limit of determination, for it cannot otherwise be determined. If we accept Kaiser's definition of the determination limit [17] with $x_0 = 3\sigma_0$, the value of the blank correction is always substantially greater than $\delta^{(0)}$, by (6.8b). It is therefore desirable to subtract any experimentally determinable value of the blank correction; the decrease in information content of the result as a consequence of an increased value of σ caused by subtracting the blank correction will always be negligible compared with the gain in information content reached by reducing or even removing the systematic error.

It follows from this discussion that, in order to obtain a high information content in the results of a quantitative analysis, it is necessary (1) to remove or at least to reduce the systematic error of the result, (2) to work under conditions for which the applied analytical method has as small a variance as possible, and (3) to evaluate the applied analytical method statistically in a series of n_s independent parallel determinations (n_s sufficiently large).

6.6   INFORMATION CONTENT OF THE RESULTS OF HIGHER-PRECISION ANALYSIS

In analyzing a sample of unknown composition, we sometimes proceed in such a way that we carry out a preliminary determination by the means of a semiquantitative method which enables to determine approximately the contents of the majority of the components present in the sample, and then make the results of individual semiquantitative determinations more precise by the use of methods that provide closer and more accurate results. If we take into account the information content of such a higher-precision analysis, the a priori distribution is given by the probability distribution of the results of the preliminary determination. The a posteriori distribution is represented by the distribution of the results after the method was made more precise.

The case when the results of both analyses are normally distributed is described by Eckschlager and Vajda in [18]. If the a priori distribution

$$p_0(x) = \frac{1}{\sigma_0\sqrt{2\pi}} \exp\left[-\frac{1}{2}\left(\frac{x-\mu_0}{\sigma_0} \right)^2 \right]$$ (6.9a)

and the a posteriori distribution

$$p(x) = \frac{1}{\sigma\sqrt{2\pi}} \exp\left[-\frac{1}{2}\left(\frac{x-\mu}{\sigma} \right)^2 \right]$$ (6.9b)

where $\sigma_0 \geqslant \sigma$, the information content is expressed by the use of the divergence measure [1, 12, 18],

$$I(p,p_0) = \ln\frac{\sigma_0}{\sigma} + \frac{1}{2}\frac{(\mu-\mu_0)^2 + \sigma^2 - \sigma_0^2}{\sigma_0^2}$$ (6.10)

If we substitute a coefficient of higher precision, $A = \sigma/\sigma_0$ $(A \leqslant 1)$, and a coefficient of result modification $B = (\mu - \mu_0)/\sigma_0$, we can simplify (6.10) to

$$I(p,p_0) = \tfrac{1}{2}(A^2 + B^2 - 1) - \ln A \qquad (6.11)$$

Details are given in references [2, 12, 18]; the finding of the information content of the results of higher-precision analyses led for the first time to the use of the divergence measure for the purposes of judging analytical results. If the information content of the results of higher precision is greater, then A and B are greater, but B is more important than A because A usually represents only a small contribution to the magnitude of the value of $I(p,p_0)$. This value gets considerably simpler in extreme cases; for example, if repeated higher-precision analysis confirms the result of the preliminary analysis (i.e., $\mu = \mu_0$), then $B = 0$ and

$$I(p,p_0) = \tfrac{1}{2}(A^2 - 1) - \ln A \qquad (6.12a)$$

In the cases $\sigma_0 \gg \sigma$ and $\mu \approx \mu_0$ (i.e., if much higher precision of results is reached but their modification is only small), then

$$I(p,p_0) = -\ln A - \tfrac{1}{2} \qquad (6.12b)$$

and finally, if no higher precision has been reached, (i.e., $\sigma_0 = \sigma$), then $A = 1$ and

$$I(p,p_0) = \tfrac{1}{2}B^2 \qquad (6.12c)$$

and the information content depends only on the degree of modification of the result.

To express information contents of results of repeated higher-precision analyses units have been introduced [18] that are defined in terms of the unit *isoinform* for corresponding pairs of values of A and B. The course of such curves connecting points with unit information content is shown in Table 6.6 and Figure 6.4. The

TABLE 6.6　Isoinform for

$$I(p,p_0)= \frac{1}{2}\left[\frac{(\mu-\mu_0)^2+\sigma^2-\sigma_0^2}{\sigma_0^2}\right] - \ln\frac{\sigma}{\sigma_0}=1$$

$\dfrac{\mu-\mu_0}{\sigma_0}$	$\dfrac{\sigma}{\sigma_0}$
0.00	0.23
0.71	0.30
1.01	0.40
1.27	0.60
1.38	0.80
1.41	1.00

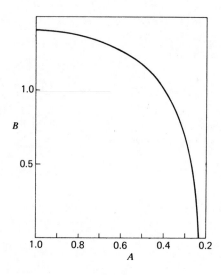

Figure 6.2　Isoinform curve for the information content of higher-precision quantitative results.

question of the units used to express information content is discussed in [12, 18].

Values of $I(p,p_0)$ according to (6.11) for various values of A and B are computed in Table 6.7. Computation of the information content by the use of the same expression will be presented in the following example. If we find by orientation analysis that $\mu_0 \approx 10\%$

TABLE 6.7 Information Content of the Results of Higher-Precision Analysis:

$$I(p,p_0) = \frac{1}{2}\left[\frac{(\mu - \mu_0)^2 + \sigma^2 - \sigma_0^2}{\sigma_0^2}\right] - \ln\frac{\sigma}{\sigma_0}$$

B	A					
	1.00	0.95	0.90	0.50	0.10	0.05
0.00	0.000	0.003	0.010	0.318	1.808	2.497
0.20	0.020	0.023	0.030	0.338	1.828	2.517
0.40	0.080	0.083	0.090	0.398	1.888	2.577
0.60	0.180	0.183	0.190	0.498	1.988	2.677
0.80	0.320	0.323	0.330	0.638	2.128	2.817
1.00	0.500	0.503	0.510	0.818	2.308	2.997
1.50	1.125	1.128	1.135	1.443	2.933	3.622
2.00	2.000	2.003	2.010	2.313	3.808	4.497
3.00	4.500	4.503	4.510	4.848	6.308	6.997

and $\sigma_0 = 5\%$ and by repeated higher-precision analysis that $\mu = 12.64\%$ and $\sigma = 0.02\%$, then $A = 0.02/5 = 4 \times 10^{-3}$ and $B = 2.64/5 = 0.528$. The information content $I(p,p_0) = 5.161$ natural information units. Here it is possible to show one more property of the information content according to (6.3) and (6.11). Prior to the orientation preliminary analysis, we know nothing more about the analyzed sample than that $0\% \leqslant X_i \leqslant 100\%$. By means of the preliminary analysis we have found that $\mu_0 \approx 10\%$ and $\sigma_0 = 5\%$. The information content of this result is $I(p,p_0) = \ln 100/5 \cdot \sqrt{2\pi e} = 1.577$ units, and that of the higher-precision analysis is 5.161 units, so the total information content $I(p,p_0) = 1.577 + 5.161 = 6.738$ units. This is somewhat lower than the information content that the result of the analysis would yield if we had analyzed the sample directly by the higher-precision method; that is, it is lower than $I(p,p_0) = \ln 100/0.02 \cdot \sqrt{2\pi e} = 7.098$ natural information units. We can, of course, conceive of a theoretical way to precisely determine a completely unknown composition in a sample, but in practical work we could not choose an appropriate analytical method without preliminary information obtained by means of a semiquantitative method. It would obviously be possible to derive an alternative form of (6.10) for other a priori and a posteriori probability distributions; but such expressions would be appropriate only in very special cases.

6.7   INFORMATION CONTENT OF TRACE-ANALYSIS RESULTS

Trace-analysis results can be of two sorts: (1) if the content of the component to be determined is $X \leqslant x_0$, where x_0 is the determination limit of the applied analytical method, we can only state that the content of the component $X \in \langle 0, x_0 \rangle$; and (2) the content of the component to be determined is $X > x_0$ and can be found quantitatively.

Since results of trace analyses are usually subject to the "shifted" log-normal probability distribution (see Section 3.3.4), we can consider the following probability densities to be the a posteriori distributions

$$p(x) = \frac{1}{x_0} \qquad\qquad x \in \langle 0, x_0 \rangle$$

$$p(x) \doteq \frac{\exp\left\{ -\frac{1}{2} \left[\dfrac{\ln(x - x_0) - \ln k x_0}{\sigma} \right]^2 \right\}}{(x - x_0)\sigma\sqrt{2\pi}} \qquad x \in (x_0, x_1)$$

$$(6.13)$$

where x_0 is the limit of determination, x_1 the highest assumed content of the trace component to be determined, and k the parameter of asymmetry of the shifted log-normal distribution. For the expected value we can find

$$E[\xi] = x_0 \left[1 + k \exp\left(\tfrac{1}{2}\sigma^2 \right) \right]$$

for $x \in (x_0, x_1)$ and

$$E[\xi] = \tfrac{1}{2} x_0$$

for $x \in \langle 0, x_0 \rangle$; the variance has been found as

$$V[\xi] = k^2 x_0^2 \exp(\sigma^2) [\exp(\sigma^2) - 1]$$

for $x \in (x_0, x_1 \rangle$ and

$$V[\xi] = \tfrac{1}{12} x_0^2$$

for $x \in \langle 0, x_0 \rangle$. The information content of trace analysis results for the case $X \leqslant x_0$ and for $x_1 > x_0$ is given by

$$I(p, p_0)_1 = \ln \frac{x_1}{x_0} \tag{6.14}$$

and for the case $X > x_0$ by

$$I(p, p_0)_2 = \ln \frac{x_1}{x_0} \frac{\sqrt{n_p}}{k\sigma\sqrt{2\pi e}} = I(p, p_0)_1 + \ln \frac{\sqrt{n_p}}{k\sigma\sqrt{2\pi e}} \tag{6.15}$$

This equation can then be used only if $0 < k < [(x_1/x_0) - 1]$ $\exp(-1.65\sigma)$. Comparing (6.14) with (6.15) we find that $I(p, p_0)_2 > I(p, p_0)_1$ if $\sqrt{n_p} / k\sigma\sqrt{2\pi e} > 1$, which is true if the expectation $E[\xi]$ is only slightly greater than x_0. Thus, the information content of a quantitative result of trace analysis is in practice always greater than the content of the finding that the amount of the component to be determined is less than the determination limit. From the analytical standpoint it is important that:

1. The determination limit of the applied analytical method has an influence on the information content of trace analysis results, in which both $I(p, p_0)_1$ according to (6.14) for $X \leqslant x_0$ and $I(p, p_0)_2$ according to (6.15) for $X > x_0$ are greater the smaller the determination limit x_0.
2. The information content $I(p, p_0)_2$ given by (6.15) for $X > x_0$ is the greater the larger the number of parallel determinations n_p which we carry out, and the more precise are the results of the

determination (i.e., the smaller is σ). The information content $I(p,p_0)_1$ for $X \leqslant x_0$ given by (6.14) does not depend on n_p and σ at all.

A practical use of the results in (6.14) and (6.15) is shown in the following example. The limit of determination of the applied analytical method $x_0 = 10^{-12}\%$ and the highest expected content of the trace component to be determined, $x_1 = 10^{-8}\%$. For the case $X_i \leqslant x_0$, the information content $I(p,p_0)_1 = \ln 10^{-8}/10^{-12} = 9.210$ natural information units. However, if we find that $E[\xi_i] = 1.9 \times 10^{-12}\%$ with precision characterized by the value $\sigma = 3$ and determine that $k = 0.01$, the information content of this determination is

$$I(p,p_0)_2 = \ln \frac{10^{-8}}{10^{-12}} \frac{1}{3 \times 0.01 \times \sqrt{2\pi e}} = 11.287 \text{ units}$$

(i.e., it is greater by more than two units than in the case when $X_i \leqslant x_0$.)

The influence of the values x_0 and σ on the information content of a quantitative result of trace analysis is illustrated in this example: in determining trace contents of Ag, Cu, and Ni in highly pure alkali salts by the method of atomic absorption spectrophotometry [19], we can decrease the determination limit

TABLE 6.8 **Information Content of the Determination of Ag, Cu, and Ni by Atomic Absorption Spectrophotometry in a Concentration 1.1** x_0

Line	Direct Determination			Determination after Extraction and Stripping		
(Å)	x_0	σ	$I(p,p_0)$	x_0	σ	$I(p,p_0)$
Ag 3280.7	5×10^{-4}	4.4	7.55	1×10^{-6}	5.2	13.60
Cu 3247.5	5×10^{-4}	4.0	7.65	5×10^{-6}	4.7	12.09
Ni 2320.0	1×10^{-3}	4.7	6.79	5×10^{-5}	5.5	9.63

by two decimal places by extraction and stripping. This is accompanied by a certain decrease of precision, of course, but in spite of that, the information content of the determination carried out after the extraction is larger, as can be seen from Table 6.8.

The problems of the information content of trace-analysis results are dealt with in [2, 19] and especially in [21], where probability distributions of trace-analysis results are also discussed for the case when X_i is only slightly greater than the limit of determination x_0 of the applied analytical method.

6.8   INFORMATION CONTENT OF TWO-DIMENSIONAL INSTRUMENTAL METHODS

In current analytical practice we encounter most frequently instrumental analytical methods that require us to carry out calibrations (see Section 2.5). The system producing information in the use of instrumental methods (i.e., an analytical device) has either a one-dimensional or a two-dimensional output (see Section 2.4). In the first case they enable us to determine, more or less selectively, one component of the analyzed sample or to prove its absence. The information content of such results can be determined as described in Sections 6.2 to 6.5 and 6.7. Instrumental methods often have two-dimensional outputs, so we determine from the position of the signal z_i which component of the analyzed sample is present, and we determine the content of the component from the intensity of the signal η_i, which takes on values $y_{i,j}(j = 1, 2, \ldots, n_p)$. Often, only quantitative determination of some components is required, so the intensities of the signals in positions known beforehand, or intensities of signals in a rank known beforehand in the entire sequence of the signals, are determined.

The information content of a result of determination of a component by means of a two-dimensional analytical method can be computed as if it were determined fully independently. For assessing the process of obtaining information in terms of a two-dimensional instrumental method we can distinguish between the real information amount, which we obtain by analysis, and a potential one, which is the maximum information amount obtainable by a certain analytical method [22]. The ratio of the former to the latter enables us to find the *information redundance* of the analytical method.

The real amount of information obtained from analysis by a two-dimensional analytical method is given as

$$M(p,p_0)_A = \sum_{i=1}^{k} I(p,p_0)_i \tag{6.16}$$

where k is the number of all simultaneously determined components. For the case of normally distributed results, we have

$$M(p,p_0)_A = \sum_{i=1}^{k} \ln \frac{(x_2 - x_1)_i \sqrt{n_p}}{\sigma_i \sqrt{2\pi e}} \tag{6.17a}$$

If the parameter σ is unknown and only its estimate s is available, then

$$M(p,p_0)_A = \sum_{i=1}^{k} \ln \frac{(x_2 - x_1)_i \sqrt{n_p}}{2s_i t(v)} \tag{6.17b}$$

where $t(v)$ is the critical value of the Student t-distribution for $\alpha = 0.038794$. If the results of a determination are subject to a mean error δ_i, then

$$M(p,p_0)_A = \sum_{i=1}^{k} \left[\ln \frac{(x_2 - x_1)_i}{\sigma_i \sqrt{2\pi e}} - \frac{1}{2} \left(\frac{\delta_i}{\sigma_i} \right)^2 \right] \tag{6.17c}$$

In case of simultaneously determining several trace components with a two-dimensional instrumental method, we get

$$M(p,p_0)_A = \sum_{i=1}^{k_1} \ln \frac{x_{1,i}}{x_{0,i}} + \sum_{j=1}^{k_2} \ln \frac{x_{1,j}}{x_{0,j}} \frac{\sqrt{n_j}}{k_j \sigma_j \sqrt{2\pi e}} \tag{6.18}$$

where k_1 is the number of components that were sought but not found, and k_2 is the number of trace components that were quantitatively determined. It is evident that $k_1 + k_2 = k$. The relation (6.16) and those derived from it [(6.17a), (6.17b), (6.17c), and (6.18)] express the real information amount obtained by instrumental analysis.

The potential information amount (i.e., the maximum we can obtain by the use of a certain analytical method), is given by (see Doerffel [9] or Danzer [23])

$$M(p,p_0)_P = \frac{z_{max} - z_{min}}{\Delta z} \log \frac{(y_{max} - y_{min})\sqrt{n_p}}{2\bar{s}_y t(\nu)} \tag{6.19}$$

where $\bar{s}_y$ is the average standard deviation characterizing the average precision of determining the intensity of a signal. Doerffel [9] has related the potential information amount to the position z_i and to the intensity of the signal considering the number of possible discriminations of the substances

$$N_1 = \frac{z_{max} - z_{min}}{\Delta z}$$

and the number of possible discriminated concentrations

$$N_2 = \frac{y_{max} - y_{min}}{2\bar{s}_y t(\nu)} \sqrt{n_p}$$

Obviously, $k \le N_1$. For the value of N_1, the minimum distance between two neighboring signals in which both signals can still be distinguished, Δz is important. For peak-shaped signals, which are very frequent in analytical chemistry, the distance depends on their profiles (Section 2.4) and on whether we take the height or the area of the peak for the intensity of the signal. These problems are discussed, for example, by Doerffel [24], Eckschlager [25], and others. For N_2, the value $\bar{s}_y$ characterizing the precision of determining the intensity of the signal is of importance. Danzer [23] states that relation (6.19) holds only if the value s_{y_i} is not dependent on the intensity y_i of the signal. However, in practice, rather the relative value $s_{r,i} = s_{y_i}/y_i$ is independent of y_i; it is then more reasonable to use the expression

$$M(p,p_0)_P = \frac{z_{max} - z_{min}}{\Delta z} \log \left[\frac{\sqrt{n_p}}{2\,\bar{s}_r t(\nu)} \ln \frac{y_{max}}{y_{min}} \right] \qquad (6.20)$$

In both (6.19) and (6.20) a general logarithm, log, is adopted; in [9] and [23] the binary logarithm ld to the base 2, which comes from communication theory, is used. For a comparison of the potential amount of information with that determined through use of the divergence measure, it is necessary to adopt the natural logarithm ln in (6.19) and (6.20).

The information redundance defined above is important primarily in considering results obtained by the use of a two-dimensional analytical method. If the redundance is generally defined as

$$\rho = \frac{M_{max} - M}{M_{max}} = 1 - \frac{M}{M_{max}} \qquad (6.21)$$

we can determine the redundance arising by the use of a two-dimensional instrumental method in such a way that we substitute $M_{max} = M(p,p_0)_P$ (the potential amount) and $M = M(p,p_0)_A$ (the actual amount), in which only one signal corresponds to each component. If, on average, m_s signals correspond to each individual component determined with a two-dimensional method, we can substitute $M = M(p,p_0)_A$, but M_{max} must be set as $M_{max} = (1/m_s)M(p,p_0)_P$. Treatises on redundance and its importance for analytical chemistry have been published by Malissa [26], Danzer [23], Doerffel [9], and others [7, 22]. Redundance is not, at least to some extent, useless: it protects our results from the possibility of gross errors, especially where several signals correspond to one component; but too large a redundance sometimes means that we cannot use a two-dimensional analytical method very effectively (see Section 6.13).

It is evident that two-dimensional analytical instrumental methods provide results with the greatest amount of information by enabling us to simultaneously determine several components of the analyzed sample (i.e., they have a great informing power). Treatises on the information properties of analytical results are often directed to the use of two-dimensional analytical methods. They have been generally discussed by Kaiser [27], Eckschlager [22, 25], and Doerffel [9]. Danzer has dealt with methods of spectral [28] and local analysis [29]; Huber and Smith [30], Palm [37], Smits et al. [31], Massart et al. [32, 33], Dupuis and Dijkstra [34], and others have dealt with chromatographic methods. Elementary analysis (i.e., simultaneous determination of C, H, and event. N in organic substances) has been studied from the point of view of information theory by Grieping et al. [35, 36] at the University of Utrecht.

6.9 AMOUNT OF INFORMATION OBTAINED BY COMBINING ANALYTICAL METHODS

To solve a given analytical problem, we need to have a certain minimum amount of information $M(p,p_0)_N$; it is evident that the problem can be solved only by means of an analytical method providing the actual information $M(p,p_0)_A \geqslant M(p,p_0)_N$. If a problem cannot be solved by the use of one analytical method, we have to combine several methods. We can take into account: (1) a *series combination* (i.e., a successive use of several methods, for example, of preliminary and repeated higher precision determinations); and (2) a *parallel combination* (i.e., the use of several methods in the same time when, for example, we determine various components of the sample with different methods).

A reason for use of the series combination may emerge from a lack of sufficiently comprehensive preliminary information, which we have to provide by a preliminary analysis. If the only preliminary information about the content of the component to be determined is that $X \in \langle x_1, x_2 \rangle$, and if the results of preliminary determinations are normally distributed with parameters μ_1 and σ_1 and the results of repeated higher-precision determinations have a normal distribution with parameters μ_2 and σ_2, the amount of information obtained by the series combination of these two methods is

$$M(p,p_0)_A = \ln\frac{x_2-x_1}{\sigma_2\sqrt{2\pi e}} + \frac{1}{2}\frac{(\mu_2-\mu_1)^2-(\sigma_1^2-\sigma_2^2)}{\sigma_1^2} \qquad (6.22)$$

Here the first term is the expression for the information content which we would obtain if we analyzed the unknown sample directly with the second higher-precision method. The second term is positive or negative according to whether $(\mu_2-\mu_1)^2 > (\sigma_1^2-\sigma_2^2)$ or $(\mu_2-\mu_1)^2 < (\sigma_1^2-\sigma_2^2)$ (see also the example in Section 6.6). From the economic point of view of processing to obtain information about the chemical composition of an analyzed sample by series combinations of analytical methods, it is expedient to use those methods for preliminary analyses which enable us to simultaneously determine the greatest number of assumed components and to obtain a low specific information price. For repeated higher-precision determinations, we use the most precise and most accurate method.

A reason for a parallel combination of methods is the case when we cannot determine all the components of the analyzed sample simultaneously by means of a single procedure. The information amount obtained in this way is given by

$$M(p,p_0)_A = \sum_{j=1}^{m}\sum_{i=1}^{k_j} I(p,p_0)_{i,j} \qquad (6.23)$$

where m is the number of methods, which we combine in the parallel way, and k_j is the number of components, which we determine by the jth method $(j=1,2,\ldots,m)$. In complicated cases we are sometimes forced to combine methods in both parallel and series ways.

6.10   INFORMATION CONTENT OF RESULTS IN ANALYTICAL QUALITY CONTROL

A frequent reason for carrying out an analysis is the quality control of a certain product. The a priori assumption of a content of the component to be determined is given here by the tolerance limits within which the content of the component is supposed to lie, yet we do not exclude in advance the chance that it may lie outside these limits. Also, analytical quality control is carried out to find out whether the content of the component to be determined lies within the prescribed limits or does not lie within them.

Tolerance limits, where x_L is the lower and x_U the upper limit of the tolerance interval, are determined on the basis of long-term experience as

$$x_L = \bar{x} - K\sigma_F$$
$$x_U = \bar{x} + K\sigma_F \tag{6.24}$$

where $\bar{x}$ is the required value of the content of the component in the product and σ_F the standard deviation characterizing the variability of the content. The tolerance coefficient K determines the width of the tolerance interval, so that with the proportion $100\gamma\%$ ($\gamma < 1$) of all products the content μ of the component lies within the interval (x_L, x_U) with the probability α. In other cases it lies outside this interval with the probability $(1 - \alpha)$. In common practical work we usually choose $0.90 \leqslant \alpha \leqslant 0.99$ and $0.80 \leqslant \gamma \leqslant 0.99$. The values of tolerance coefficients are tabulated, for example, in [38].

To express the divergence measure of the information content in analytical quality control it is important how we characterize the a priori probability distribution under the assumption that the tolerance limits x_L and x_U are given beforehand. Two of possible ways are outlined in [13]; here we will show the simpler of the two, and assume that the content of the component found by the analysis is

$X \in \langle x_L, x_U \rangle$ with probability α, and outside this interval with probability $(1 - \alpha)$. We then have

$$p_0(x) = \begin{cases} \dfrac{\alpha}{x_U - x_L} & x \in \langle x_L, x_U \rangle \\[3mm] \dfrac{1 - \alpha}{100 - (x_U - x_L)} & x \notin \langle x_L, x_U \rangle, \ 0 \leqslant x \leqslant 100 \end{cases}$$

(6.25)

Such a distribution is illustrated in Figure 6.5. The a posteriori probability distribution of the analysis results is mostly normal with parameters μ and σ; here $\sigma \leqslant \sigma_F$ always. The information content for the case when the content of the component to be determined lies within the tolerance limits (i.e., when the product meets the requirements) is given by

$$I(p, p_0)_\alpha = \ln \frac{x_U - x_L}{\alpha \sigma \sqrt{2\pi e}} \qquad \text{for } \mu \in \langle x_L, x_U \rangle \qquad (6.26a)$$

and for the case when the product does not meet the requirements because the content of the component lies beyond the tolerance limits, it is given by

$$I(p, p_0)_{1-\alpha} = \ln \frac{100 - (x_U - x_L)}{(1 - \alpha)\sigma \sqrt{2\pi e}} \qquad \text{for } \mu \notin \langle x_L, x_U \rangle \qquad (6.26b)$$

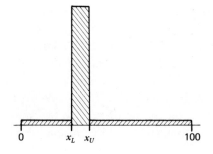

Figure 6.5 A priori distribution in quality control analysis.

TABLE 6.9 Information Content of Analytical Quality Control:

$$I(p,p_0)=\ln\frac{1}{\sigma\sqrt{2\pi e}} - \left[\phi\ln\frac{1-\alpha}{100} + (1-\phi)\ln\frac{\alpha}{x_U - x_L}\right]$$

$x_U - x_L = 1;\ \phi = \int_{-\infty}^{x_L} p(x)\,dx;\ \alpha = 0.90,\ 0.95,\ \text{and } 0.99;\ \sigma = 0.10,\ 0.05,\ \text{and } 0.01$

| | σ = 0.10 | | | σ = 0.05 | | | σ = 0.01 | | |
| | α | | | α | | | α | | |
μ	0.90	0.95	0.99	0.90	0.95	0.99	0.90	0.95	0.99
$x_L - 8\sigma$	7.791	8.484	10.094	8.484	9.178	10.787	10.094	10.787	12.396
$x_L - 2\sigma$	7.636	8.312	9.884	8.329	9.006	10.577	9.939	10.615	12.187
$x_L - \sigma$	6.712	7.786	8.634	7.405	7.979	9.327	9.014	9.589	10.936
x_U	4.390	4.710	5.493	5.083	5.403	6.187	6.692	7.012	7.796
$x_U + \sigma$	2.068	2.133	2.354	2.762	2.826	3.047	4.371	4.436	4.656
$x_U + 2\sigma$	1.144	1.107	1.103	1.837	1.800	1.797	3.447	3.410	3.406
$\frac{1}{2}(x_L + x_U)$	0.989	0.934	0.894	1.682	1.628	1.587	3.292	3.237	3.196
Difference	6.802	7.550	9.200	6.802	7.550	9.200	6.802	7.550	9.200

Since $0.90 \leqslant \alpha \leqslant 0.99$ and $(x_U - x_L) \ll 100$, $I(p,p_0)_{1-\alpha}$ is always considerably larger than $I(p,p_0)_\alpha$. It is obvious also that the pragmatic importance of finding an inconvenient amount of the product is greater than a statement expected in advance that the product has the assumed content of the controlled component. The values of $I(p,p_0)$ for various cases of results of analytical quality controls $\mu \notin \langle x_L, x_U \rangle$ and $\mu \in \langle x_L, x_U \rangle$, and for various values of σ and several values of α, are computed in Table 6.9.

It is also appropriate here to mention the minimum information content which has to be shown by the results of an analytical method used for the needs of quality control. The width of the tolerance interval $(x_U - x_L) = 2K\sigma_F$; at the same time, the condition from Section 6.3 that $(x_U - x_L) \geqslant 6\sigma$ must be fulfilled. This results not only in the condition that for analytical quality control of a product whose variability is expressed by the value σ_F, only such an analytical method may be used, for the results of which $\sigma \leqslant \frac{1}{3} K\sigma_F$, but also in the following requirement of the minimum information content:

$$I(p,p_0)_{\min} = \ln \frac{2K\sqrt{n_p}}{\alpha\sqrt{2\pi e}} = \ln \frac{\sigma}{\sigma_F} \frac{6\sqrt{n_p}}{\alpha\sqrt{2\pi e}} \tag{6.27}$$

where n_p is the number of parallel determinations $(n_p \geqslant 2)$, K the tolerance coefficient at the probability α, σ the standard deviation of the analytical results, and σ_F the standard deviation characterizing the variability of production. The minimum information content has been discussed in [3] and [13].

6.11   ASSESSMENT OF ANALYTICAL METHODS

In practical work we assess analytical methods as a rule according to what results they provide and how rapidly and with what costs these results are obtained. A method can be assessed either absolutely (i.e., as itself and eventually compared with other methods), or relatively (i.e., it is viewed from the standpoint of its suitability as applied to a concrete analytical task). Analytical results can be assessed by their information contents or by the amount of information they provide, as was shown in Sections 6.2, 6.3, and 6.6 to 6.8; in evaluating methods we also have to take into account their temporal and economic aspects.

The *temporal aspect* in the evaluation of analytical methods by the use of quantities of information theory was introduced by Doerffel [9] and expanded later by other authors [39, 40]. The amount of information transferred in a specified unit of time is usually denoted as the *information flow*:

$$J(p,p_0) = \frac{dM(p,p_0)}{dt} \qquad (6.28a)$$

This quantity is suitable for assessing permanently operating analytical methods; it is enumerated in information units $\times s^{-1}$. For the evaluation of discontinuously operating analytical procedures the *information performance* is more appropriate. It is defined as

$$L(p,p_0) = \frac{1}{t_A} \int_0^{t_A} J(p,p_0)\,dt = \frac{1}{t_A} M(p,p_0) \qquad (6.28b)$$

where t_A is the time of duration of the analysis. Since the time necessary for the analysis to be carried out can be split into the time t_0 independent of the number of parallel determinations and the time t_n dependent on its number (e.g., as $t_A = t_0 + n_p t_n$; see Section 2.7), the information performance is a quantity suitable, for example, in considerations of optimum numbers of parallel determinations.

Economic aspects of assessing analytical methods [3, 39, 41] are also used frequently. As a rule the *specific information price* is defined as

$$C(p,p_0) = \frac{\tau}{M(p,p_0)} \qquad (6.29)$$

where τ is the cost of carrying out the analysis. The specific information price is enumerated in ratios of currency units and information units (e.g., in $/bit). Since the economic time equivalent can be included in the cost τ, the specific information price can be employed more generally than, for example, information flow or information performance. Then, of course, not only must the wages of analysts be included in the cost (the active time of analysts must be separated from their passive time, during which procedures run automatically) but also the economic loss that arises from waiting for the results of the analysis. Such quantities as the information flow or the information performance and the specific information price enable absolute assessment of analytical methods and their mutual comparison.

In common analytical practice relative evaluation of analytical methods (i.e., assessment of suitability of their use in a concrete analytical problem) is more important. A quantity appropriate for relative assessment of an analytical method is the *information profitability* [4], defined as

$$P(p,p_0) = \frac{E}{\tau} M(p,p_0) \tag{6.30}$$

where the effectivity E can be defined in different ways but must always express the mutual relation between the properties of the results required for the given task and those provided by the analytical method. One of possible ways to define the effectivity is Danzer's definition, which is strict but suitable for some purposes [4, 40]:

$$E = \prod_{i=1}^{n_\varepsilon} \varepsilon_i \tag{6.31a}$$

E is the product of n_ε effectivity coefficients for which we substitute

$$\varepsilon_i = \begin{cases} e_i & e_i \leqslant 1 \\ 0 & e_i > 1 \end{cases} \tag{6.31b}$$

where, for instance, the coefficient of precision $e_1 = \log \sigma_A / \log \sigma_R,$

with σ_R being the required and σ_A the actual standard deviation, and the coefficient of time effectivity $e_2 = t_A / t_R$, with t_R being the time interval within which we require to know the result. Similarly, we can define $e_3 = k_R / k_A$, where k_R is the number of components of an analytical sample we need to determine and k_A the number of components that can be determined simultaneously by a particular analytical method. In evaluating trace-analysis methods, $e_4 = \log x_{0,A} / \log x_{0,R}$ may take place, with $x_{0,R}$ being the required and $x_{0,A}$ the actual limit of determination. In assessing microprobe methods, the coefficient $e_5 = A_R / A_P$ may be useful, where A_R is the needed and A_P the determined surface discriminating capability (see Section 2.3); and so on. Basically, we can introduce effectivity coefficients for any property to be assessed that can be expressed quantitatively.

It follows from (6.31a) and (6.31b) that $E = 0$ if only one coefficient $e_i > 1$, which implies that the analytical method is not applicable for the given objective and the concept "effectivity" loses its sense. The importance of the information profitability given by (6.30) is that it can be well adopted as a quantity enabling us to evaluate analytical methods relatively, in the choice of the optimum analytical method and in optimizing the analytical procedure itself, in decisions on optimum instrument equipment, in considering whether we should introduce automatic data processing of analytical data in a certain case, and so on.

6.12   USEFUL NUMBER OF REPEATED DETERMINATIONS

The information content of a quantitative analysis expressed in terms of the divergence measure by (6.5) depends on the number n_p of parallel determinations, on the number n_s of determinations from which the standard deviation s has been estimated, and, in methods for which a calibration curve is constructed, on the number N of points which this curve fits. This dependence is such that the information content $I(p, p_0)$ increases with increasing values of n_p, n_s, and N. Practical problems connected with carrying out a large number of repeated determinations arise primarily because such determinations prolong the duration of the analysis. We will therefore assess the expedient number of repeated determinations by means of the information performance [3]:

$$L(p,p_0) = \frac{1}{t_A} I(p,p_0) \qquad (6.32a)$$

or for the results of an analysis carried out by a two-dimensional analytical method by

$$L(p,p_0) = \frac{1}{t_A} M(p,p_0) \qquad (6.32b)$$

where $M(p,p_0)$ is the actual information amount obtained by the analysis, as defined in Section 6.8.

The values of n_s or N influence the constant time t_0 in the relation $t_A = t_0 + n_p t_n$ (see section 6.11), but not significantly if the standard deviation is estimated and the construction of the calibration curve is controlled "once and for all" or only from time to time. It will therefore be purposeful to estimate s and to construct the calibration curve from sufficiently large values of n_s and N, as was shown in Section 6.4, and to determine an expedient number n_p of parallel determinations, in which $n_p \geqslant 2$, so that control of the reliability of the results be possible, for example, by comparing their range with the value of the allowed difference of the parallel determinations. Since $I(p,p_0)$ or $M(p,p_0)$ increase with the number of parallel determinations, at first rapidly for small values of n_p and then less rapidly for higher values of n_p, but $1/t_A = 1/(t_0 + n_p t_n)$ decreases with n_p more or less rapidly according to the ratio of t_0 and t_n, it is obvious that $L(p,p_0)$ either decreases or takes on its maximum value with increasing values of n_p for $n_p \geqslant 2$. Where the information performance $L(p,p_0)$ has its maximum value for $n_p \leqslant 2$, we take $n_p = 2$ as the optimum value and in other cases we take that value of n_p for which $L(p,p_0)$ is maximum; in practical work, $2 \leqslant n_p \leqslant 5$. Evidently, more than two parallel determinations appear as optimum only when the number of parallel determinations carried out influences but little the time of the duration of the analysis t_A. The establishment of such a number of parallel determinations, for which the information performance of a method is maximum, will be presented in the following two examples.

If we have for a standard determination $x_2 - x_1 = 10$, $s = 0.012$, and $t(\nu) = 2.1811$ for $n_s = 26$, and the time of duration of the analysis $t_A = 18 + 7n_p$ minutes, we get the information performance

$$L(p,p_0) = \tfrac{1}{25} \ln \frac{10\sqrt{1}}{4.362 \times 0.012} = 0.210 \text{ natural unit/min}$$

for $n_p = 1$,

$$L(p,p_0) = \tfrac{1}{32} \ln \frac{10\sqrt{2}}{4.362 \times 0.012} = 0.175 \text{ natural unit/min}$$

for $n_p = 2$, and

$$L(p,p_0) = \tfrac{1}{39} \ln \frac{10\sqrt{3}}{4.362 \times 0.012} = 0.149 \text{ natural unit/min}$$

for $n_p = 3$. The method shows its maximum information performance at $n_p = 1$; hence, in practical work with the analytical method we will use $n_p = 2$ as the minimum number of parallel determinations with which reliability control of the results is possible.

Let us propose, on the other hand, an instrumental analytical method with $x_2 - x_1 = 10$, $t(\nu) = 2.2117$ for $n_s = 21$, the time of duration of the analysis $t_A = 43 + 2n_p$ minutes, and the standard deviation of determinations carried out by the use of a calibration curve for $N = 8$ given as

$$s = 0.0075 \sqrt{\frac{1}{n_p} + 0.125}$$

Here the information performance $L(p,p_0) = 0.126$ information unit/min for $n_p = 1$, $L(p,p_0) = 0.134$ information unit/min for $n_p = 2$, $L(p,p_0) = 0.136$ information unit/min for $n_p = 3$, $L(p,p_0) = 0.135$ information unit/min for $n_p = 4$, and $L(p,p_0) = 0.133$ information unit/min for $n_p = 5$. Hence, the optimum number of parallel determinations in the application of this analytical method is $n_p = 3$.

Similar approach is inapplicable when carrying out trace analyses, where we often try to improve the signal-to-noise ratio by performing a higher number of parallel determinations [44, 45], and also in the case of permanently and automatically operating analytical methods.

6.13   CHOICE OF THE OPTIMUM ANALYTICAL METHOD

The optimization of the process of obtaining information about the chemical composition of the matter can be split into two basic problems: (1) the *choice of the most suitable method* for the given problem from a number of analytical methods applicable to the answer to it, and (2) the *choice of conditions* under which the selected analytical method provides the most appropriate results in the given problem. In both cases the information profitability according to (6.30) (i.e., a quantity assessing the method or the results relatively) can serve as a criterion for reaching the optimum value.

The choice of the most suitable analytical method is usually simpler than the optimization of conditions, and is often determined simply by the instruments available in the laboratory. Yet we sometimes have to choose such a method from a number of possibilities; then the information profitability according to (6.30) with the effectivity defined by (6.31a) and (6.31b) fits as a criterion of the suitability of a particular method. It is evident that we choose that method which has the greatest information profitability from a set of methods applicable to answer the specific problem. In adopting this quantity as a criterion for the choice of the most suitable analytical method, it is important to know precisely the requirements that will be imposed on its results when the method is used, as only a small change in the requirements can often result in the choice of a method other than the one we would choose based on the initial requirements. Requirements must not be established in such a way that we require the greatest precision, the shortest time of the analysis, and/or the greatest number of simultaneously determined components, where there are no rational reasons for these. Such nonrealistic demands upon the properties of a test method result nearly always in a disproportionally high cost and therefore in a decrease of profitability; on the other hand, little pretentious requirements (e.g., upon the precision) result in too little information; besides, for the case of a rather narrow, rectangular a priori distribution (e.g., in analytical quality control or clinical analysis), it is possible to determine the minimum information content that the analytical method must provide to be applicable for the specific purpose [3].

Choice of the most suitable analytical method in terms of information profitability is more expedient than is choice of the method that has maximum effectiveness; it sometimes proves that the method with the highest effectiveness need not be the most appropriate from the economic point of view and therefore hasn't the greatest profitability. This can be shown in an example of the choice of the most suitable method to determine 0.05 till 6% Mn in low-alloy steel [4] if the results are needed to be precise with a relative standard deviation $s_r = 0.10$ and the analysis is to last not longer than (a) 1440 minutes (1 day), (b) 30 minutes, or (c) 2 minutes. Table 6.10 presents a useful survey for decision making. In case (a), methods 1, 2, and 3 have the same profitability, which is higher than that of spectral methods, although they have different effectivities. The most effective of the spectral methods is atomic absorption spectrophotometry; it is almost as profitable as optical emission spectrography, but it has an effectivity of higher decimal order. In case (a) we would choose one of methods 1, 2,

TABLE 6.10 Comparison of Analytical Methods for Determination of Manganese in Low-Alloy Steels

Analytical Method	Relative Standard Deviation, s_r	Number of Determinable Components, N	Time, t_A (min)	Relative Cost (currency units)	Amount of Information, $M(p,p_0)$ (binary units)	Information Efficiency, E	Information Profitability, $P(p,p_0)$ (binary units/currency units)
1. Titrimetry (persulphate–silver nitrate)	0.02	1	60	16	4.00	(a) 0.018 (b) 0 (c) 0	0.0044 0 0
2. Potentiometric titration	0.02	1	30	8	4.00	(a) 0.088 (b) 0.42 (c) 0	0.0044 0.21 0
3. Photometry (permanganate)	0.03	1	30	8	3.41	(a) 0.010 (b) 0.49 (c) 0	0.0044 0.21 0
4. Atomic absorption spectral photometry	0.01	1	20	6	5.00	(a) 0.0047 (b) 0.22 (c) 0	0.0039 0.19 0
5. Optical emission spectroscopy	0.05	50	3	1	133.8	(a) 0.000026 (b) 0.0013 (c) 0	0.0035 0.17 0
6. Optical emission spectrography	0.1	50	30	12	83.8	(a) 0.0042 (b) 0.020 (c) 0	0.0029 0.14 0
7. Optical emission spectrometry	0.05	50	1	20	133.8	(a) 0.0000087 (b) 0.00042 (c) 0.0063	0.000058 0.0028 0.042

and 3 as the most suitable (i.e., one of the classic analytical methods). For case (b) we would probably choose method 3, photometric determination of permanganate, or method 2, potenciometric titration, or method 4, atomic absorption spectrophotometry. The methods of optical emission spectrography and spectrometry have lower profitabilities. In case (c), when the result is required within 2 minutes, the only applicable method is optical emission spectrometry performed by means of a quantometer. The use of this method is not too profitable because of the high relative cost, but the higher cost spent, for example, in the control of the content of Mn in the steel in iron works is offset by the decrease of economic loss caused by waiting for the results of the analysis.

The choice of analytical method or the optimization of an analytical procedure is often made with regard to the relevance of obtained information. We explained above that the relevance of empirically obtained information is given by the degree to which this information contributes to recognition of the investigated action or object. In analytical practice the relevance of information about the presence or the content of a certain component in the analyzed sample is given by the usefulness (importance, significance) of the information for the assessment of the analyzed sample and depends, primarily, on the objective of the analysis. If we obtain by a single procedure (e.g., by the use of an instrumental method) information about the content of several components (information about the contents of various components having different relevance), we optimize the whole procedure in such a way that the information with the highest relevance can be obtained with possibly the greatest information content (i.e., such that the precision of the determinations of the most important components for the assessment of the analyzed sample be as great as possible). Irrelevant results obtained simultaneously with relevant ones in an analytical procedure are usually not even evaluated in practical work.

It is obvious that analytical information that is irrelevant in one stage of investigation can become highly relevant in a later stage of the investigation; or seemingly relevant information may later lose its relevance. This results in a change or the introduction of new analytical methodology. For an objective choice of optimum analytical methodology from the standpoint of the relevance of the results, it is disadvantageous to assess the information relevance intuitively and only qualitatively; eventually, we often form a relevance ranking.

6.14   OPTIMIZATION OF ANALYTICAL PROCEDURES

The great number of factors that influence the results of an analysis and their properties, and the lack of lucidity of the influence of individual factors and their mutual interactions, force us to carry out statistical optimization, that is, seek for conditions under which the method provides the most appropriate results for a given purpose. We are nearly always concerned with seeking for extreme values of an objective function, which is dependent on many factors, often in interaction. A slight advantage is the fact that the optimums of analytical methods are usually flat and that we can better speak of an area than of a point at which a maximum is practically reached.

As the most appropriate methods used to find the region of conditions under which a process shows optimum properties (i.e., an extreme value of the objective function) appear, the sequence simplex method introduced by Spendley et al. [77] and the method of maximum slope of Box and Wilson [78]. The application of the simplex method in analytical chemistry has been suggested by Deming and Morgan [63, 64]. They also showed its practical application [64]. Doerffel et al. [24, 65] have used the method of maximum slope and have shown in [11] that the simplex method can result in finding a side maximum. The simplex method is described in Section 5.6. In common analytical practice the choice of a suitable objective function appears to be a greater problem; this is explicitly noted, for example, by Smits et al. [31]. Objective functions based on information quantities have been used on a number of occasions in recent times. The function used is always either the information performance (the specific information price could be adopted as well) or the information profitability.

In optimizing any process, including an analytical one, the choice of a suitable *objective function* is always very important. Such an objective function has to include in an appropriate quantitative form all the properties we want to optimize. It is obvious that a process running in an optimum way with respect to an objective function may appear to be suboptimal for another objective function. In practice, the choice of the suitable objective function is the basis of all successful optimizations of analytical procedures. Optimization is carried out today mainly in the procedures of instrumental analysis, as, for example, in separation methods (chromatography) or spectral methods, whose results, properties, and temporal/economic aspects of use are influenced by a large number of factors. Ways of seeking an optimum have been given above; here we want to show the possibilities of applying information quantities as objective functions.

In most simple cases the information performance $L(p,p_0)$ by (6.28b) can be used; in Section 6.12 this quantity was adopted for seeking an optimum number of parallel determinations. Here, of course, we were not concerned with optimizing an analytical process in the common sense. Similarly, we could use the specific information price, which we would minimize, as the objective function in some simple cases. Doerffel [11], who noted the possibilities of optimizing information properties of analytical methods in 1973, has used the quantity

$$C = \frac{N_1}{t_A} \log_2 \frac{(x_2 - x_1)\sqrt{n_p}}{\sigma\sqrt{2\pi e}} \tag{6.33}$$

as the objective function, where $N_1 = (z_{max} - z_{min})/\Delta z$ is the number of possible discriminated substances (see Section 6.8), t_A is the time of duration of the analysis, and the rest of (6.33) is, in fact, the information content according to (6.3) but expressed in binary logarithms. He illustrates the use of the quantity C in an example of optimizing the process of the spectrometric solution analysis with a stabilized arch and optimizing the time-dependent intensity of the signal in the qualitative proof of silicon by use of the same methodology. In another paper [65] he has dealt with optimizing the process of spectroanalysis with a stabilized arch and of gel chromatography. Smits et al. [31] have optimized the process of separating Cd, Zn, Cu, Mn, and Ca in a cation-exchange column by the simplex method, using a quantity they denote as the *informing power* and define by

$$P_{\text{inf}} = \frac{\sum \log_2 \dfrac{1}{\Omega_{i-1,i} + \Omega_{i,i+1}}}{t} \tag{6.34}$$

where t is the time needed to completely elude the component with the highest elution coefficient and $\Omega_{i-1,i}$ is the overlap between the $(i-1)$th and ith peaks. Thus, it is a quantity having the nature of information performance if the content is expressed in terms of the Brillouin measure [79] (i.e., by means of the number of binary discriminations).

Smits et al. have drawn attention [31] to the fact that in optimizing an analytical procedure, where we are always concerned with multifactor experimental optimization, the most difficult task is to find a suitable objective function; the application itself (e.g., of a simplex method) is usually easier. Similar experience was gained by Eckschlager [61], who optimized the process of emission spectrography in determining minor and trace contents of metals by the use of the information profitability (6.30), which is a quantity confirming the response relatively [in contrast to the information performance of the Doerffel's quantity C by (6.33) or Smits' P_{inf} by (6.34)], that is, with respect to the properties of the results required for the answer to the given analytical problem. The information profitability is written for optimization needs as

$$P(p,p_0) = \frac{E}{\tau} \left\{ \sum_{i=1}^{k_1} \ln \frac{(x_2 - x_1)_i \sqrt{n_p}}{\sigma_i \sqrt{2\pi e}} + \sum_{j=1}^{k_2} \ln \frac{x_{1,j}}{x_{0,j}} \right.$$
$$\left. + \sum_{l=1}^{k_3} \ln \frac{x_{1,l}}{x_{0,l}} \frac{\sqrt{n_p}}{k_l \sigma_l \sqrt{2\pi e}} \right\} \tag{6.35}$$

where k_1 is the number of determined macrocomponents, k_2 the number of trace components sought for but not found, and k_3 the number of trace components found and quantitatively determined. In [61] two alternatives are taken into account: (1) to optimize the procedure in such a way that determinations of the macrocomponents and the trace components both have maximum profitability, that is, to look for compromise conditions, and (2) to determine macrocomponents and trace components separately and to optimize both the process of determining the macrocomponents (the maximum precision and accuracy) and the process of determining the trace components (the minimum determination limit). It is shown that sometimes alternative 1, at other times alternative 2, is more advantageous; here much depends on what requirements are set beforehand, especially as far as the values of σ_R and $x_{0,R}$ are concerned. The optimization of analytical procedures has not yet been commonly applied so that we may draw some generally valid conclusions, but from the few known examples we can at least roughly conclude that (1) it is very useful to optimize analytical procedures, primarily in applications of instrumental methods in routine control, clinical, and similar activities; (2) the greatest difficulty in optimizing an analytical procedure is the choice of a suitable objective function; and (3) information quantities such as information performance that evaluate analytical methods absolutely, and quantities such as information profitability that evaluate analytical methods relatively, can be adopted as appropriate response functions in optimizing analytical procedures.

6.15 METHODS OF PATTERN RECOGNITION

Methods of pattern recognition have been used recently and successfully for evaluation and interpretation of analytical information, for classification and identification purposes, and for the choice of an optimum analytical method. These, originally empirical methods, can be based today on the concepts defined in information theory (cf. K. Varmuza [81] or H. Rotter and K. Varmuza [82]). The principal advantage of the pattern recognition methods is the possibility of employing a large number of parameters for resolutions that can be expressed by digital descriptors. The correctness of a resolution is not threatened even if we use some parameters that later appear as irrelevant for the resolution. The importance of the choice of an analytical method with respect to a large number of parameters was pointed out by Kaiser [27].

As *pattern recognition* those methods are recognized that enable to analyze multivariate experimental data and classify objects, characterized by these data, into classes defined by some patterns according to similarity of the objects in pursued properties. Sometimes it is possible in addition to use methods of pattern recognition for testing and starting working hypotheses. The methods of pattern recognition have found their assertion in chemistry already earlier [83]; analytical results are suitable for being evaluated with these methods especially because modern analytical methods enable to acquire relatively easily a sufficient amount of data (spectra, chromatograms, results) with suiting precision.

The algorithms for categorizing objects on the basis of data measured on them arise always in four steps:

1. The physical world is sensed by some transducer system which inputs its data into the *pattern space*. The transducers describe a representation of the infinite dimensionality of the parameters of the real world by a (large) number of scalar values. This number becomes then the dimensionality of the pattern space. A column vector x in the pattern space will have scalar elements

$$\mathbf{x} = (x_1, x_2 \ldots, x_R)^T$$

where each x_i $(i = 1, 2, \ldots, R)$ represents the particular value associated with the ith dimension. Thus i indexes the dimensions of the pattern space and T signifies "transpose". The vector $\mathbf{x}$ can be interpreted as a point in R-dimensional space with coordinate values x_i.

2. The *feature selection*, that is, finding a minimum number of relevant variables for a given classification, is an intermediate

domain between the data gathering space and the classification process. It is the most pretentious part for the human intellectual activity. The need for a feature space lies in the requirement of a space in which classification algorithms can be efficiently implemented. Even the simplest algorithms are quite time-consuming on large scale digital computers and the dimensionality of the pattern space can be quite high. It is therefore desirable to reduce the dimension R to a much smaller dimension N, yet maintaining the discriminatory power for classification purposes.

One technique for selecting features is searching for maximization of dissimilarity between classes by the use of the concept of entropy borrowed from information theory. Thus, if we consider a source of classes S_k,

$$H(S) = - \sum_{k=1}^{K} P(S_k) \ln P(S_k)$$

becomes, according to (4.12), the source entropy (enumerated in natural information units). If we have quantitative information $q_r(j)$ on the rth dimension of our R-dimensional pattern space, we want a measure, to select the "best" feature. This implies the need for a conditional probability $P(S_k | q_r)$ and the entropy of the source given an observed feature value becomes

$$H(S|q_r) = - \sum_k \sum_j P(S_k | q_r(j)) \ln P(S_k | q_r(j))$$

This is known as the equivocation of the source given $q_r(j)$. We are interested in learning as much about the classes by observing the fewest (but most information-bearing) dimensions in order to achieve a meaningful feature selection. The information amount gained by observing a certain dimension must be

$$I(S, q_r) = H(S) - H(S|q_r)$$

which means that the uncertainty of the source has been reduced by the amount of the equivocation. Thus this quantity, known as the mutual information of the rth dimension taking on the value

q_r, is the criterion of "goodness" of that particular coordinate. By ordering the coordinate dimensions such that

$$I(S, q_1) \geqslant I(S, q_2) \geqslant \cdots \geqslant I(S, q_R)$$

and retaining only the N largest dimensions, the pattern space will be reduced to the feature space.

   In spite of conditioning the success of the entire classification, the feature selection is not yet based on a general theoretical principle; however, it is true that omitting a relevant feature cannot be improved by any mathematical approach while an irrelevant feature is automatically set aside during the process. It is therefore expedient to start the pattern recognition with more features including those whose relevance may be doubtful. Vandeginste [84], for example, has chosen an analytical method for some purposes by the use of ten features, some of which are irrelevant in some cases but are of great importance in other cases (e.g., the value of σ in quantitative and qualitative determinations, the determination limit in determining main and trace components, etc.). The features can be quantitative or qualitative variables, the latter ones being coded by a descriptor, most frequently a binary one. In [81] and [82] the use of information content in feature selection has been dealt with. A two-step feature selection has been studied by Štrouf and Fusek [85] who have introduced the concept of "intrinsic dimensionality" for the least number of relevant features.

3. The *classification* of objects into classes defined by various patterns according to feature similarities can be carried out in several ways. We often have a priori knowledge of some data vectors in the pattern space as to their correct classification. These prototypes can be denoted as column vectors

$$\mathbf{y}_m^{(k)} = \left(y_{1m}^{(k)}, y_{2m}^{(k)}, \ldots, y_{Rm}^{(k)} \right)^T$$

where k indexes the particular class and m indicates the mth prototype of class S_k. The classification problem lies now in finding separating surfaces in R-dimensional space that correctly classify the known prototypes and that, according to a criterion, will correctly specify unknown patterns. In order to carry out such a task successfully, we have to define similarity measures between points in our R-dimensional pattern space.

The most frequent are the distribution free methods based either on discrimination functions or on suitably chosen measures of similarity. A discrimination function can be linear or nonlinear. The similarity measures can be based on distances defined in different ways [81, 82, 86, 90] or on assessing of fitting regression functions as has been done in SIMCA (Statistical Isolinear Multiple Component Analysis), introduced by Wold [87]. The classification itself, adopting an algorithm, is processed in a computer.

4. The last part of a pattern recognition method is recording and illustrating the results.

The use of pattern recognition methods in chemical analysis has become quite ample and successful; most frequently, a complex analytical signal, mainly in mass and molecular spectrometry, is being processed [81, 82, 88, 89, 90]. Varmuza [91] has compared the identification of organic substances carried out on the basis of mass spectra by looking into a spectra library and by a pattern recognition method. After introducing standard descriptors the pattern recognition can be used for the purposes of the structural analysis (cf. Brugger et al. [92] and Hodes [93]). Pattern recognition methods can be employed for the choice of an optimum analytical method or procedure; thus, for example, Eskes et al. [94] or Massart and De Clercq [95] have used the pattern recognition for the choice of dissolvents and a stationary phase in chromatography.

6.16   OTHER USES OF INFORMATION THEORY IN ANALYTICAL CHEMISTRY

General and theoretical questions of analytical chemistry today are often answered by the use of concepts and relations of information theory. The recent literature cites applications of this theory to special and concrete cases of evaluating analytical methods from the point of view of their possibilities for answering a given analytical task, in comparing several methods, and in optimizing analytical procedures.

Let us first note general problems of analytical chemistry as they are answered by the use of information theory. Kaiser answered some questions by the use of concepts and methods of information theory in his general paper on quantitation in analytical chemistry [27] written in 1970. General problems and aspects of analytical chemistry in concepts of systems and information theories were defined by Malissa [5, 26, 43–48], Kienitz [49], and Gottschalk [42, 50]. Later, other authors [12] defined analysis as a process of obtaining information. An up-to-date definition of analytical chemistry, based on the concept of information, has been presented by Malissa [48]. Gottschalk [42, 50] has drawn attention to the possibilities of the use of systems theory, the theory of games, and information theory in analytical chemistry. At present the members of the "Lindau circle" are systematically concerning themselves with general problems of analytical chemistry, and they have used aspects, concepts, and relations taken from information theory in some of their publications [26, 42].

As far as mathematical tools are concerned, the use of measures and concepts borrowed from communication theory is obvious in earlier publications, especially in papers by Rackow [51], the pioneering work of Doerffel and Hildebrand [9], and some early papers by Eckschlager [7, 16, 39]. Imperfection of the measures of communication theory for the needs of analytical chemistry soon became evident, and new ways were sought in which to express the information properties of analytical results and methods. Danzer [6, 23, 28, 29, 40, 54] has based a whole system of evaluating analytical results and methods on the concept of information content as the binary logarithm of the number of possible discriminations, in which, similarly to the approach of Doerffel [9, 11], he has distinguished the number of possible discriminations of substances from the number of possible discriminations of concentrations. For the exact mathematical definition of basic concepts, papers by Malissa and Rendel [26] and by Gottschalk [42, 50] are of fundamental importance. The use of the divergence measure first appeared in a paper by Eckschlager and Vajda [18] in 1974; this measure, which appears today to be very suitable for assessing the information content of experimental data, was later employed in [2, 3, 8, 12, 14, 21, 22, 55, 56] and taken over by Danzer [4, 10]. Various measures of information content were compared by Eckschlager in [12]. The mutual connection between various information measures has been studied in a paper by I. Vajda and K. Eckschlager [8]. Potential and actual information contents of the results of instrumental analytical methods are defined by Danzer [28] and discussed by Eckschlager [22].

Zettler [57] and Eckschlager [58] are concerned with problems of the flow of the matter and information in obtaining analytical results; Seifert [59] has utilized semantic aspects of analytical information to answer questions of the amount of information needed and at the same time called attention to the expediency of possibly continuous information flow in production control. Temporal and economic aspects of the evaluation of analytical methods were introduced primarily by Doerffel [9, 11], Danzer [40], Huber and Smith [30], Triebel and Wilhelmi [60], Gottschalk [50], Liteanu and Crisan [41], and Eckschlager [3, 61, 62]. Information effectivity is defined by Eckschlager [3, 39, 61], Danzer [40], and Danzer and Eckschlager [41]. Optimization of analytical methods and procedures by the use of the quantities of information theory was first claimed by Doerffel [11]; Smits et al. [31] employed a quantity of the nature of information performance as an objective function to optimize separation in a cation-exchange column. A survey of the scope of use of information theory in analytical chemistry can be found in [62].

Statistical optimization of analytical results and methods appears to be quite common today and information theory is used more and more frequently to evaluate and optimize various analytical methods and procedures as applied to concrete cases.

Deming and Morgan [63, 64] have devoted two papers of basic importance to the problems of optimizing analytical methods, and Doerffel et al. [11, 65] have discussed optimization of the information properties of analytical methods. Optimization of analytical procedures is not yet common; considerably richer is the literature dealing with the comparison of analytical methods [66], primarily with evaluating analytical results and methods by the use of quantities defined in information theory. For instance, the evaluation of results and methods of qualitative and instrumental analyses has been dealt with by Eckschlager [2, 16]. Malissa [5] has compared the use of a qualitative analytical procedure and of a spot test according to the information content of the results, and Danzer [6] later discussed the amount of information obtained by the use of various procedures of qualitative analysis. Fischer [67] called attention to the relatively high information content of employing test papers for semiquantitative examinations in clinical diagnostics (e.g., pH, albumin, blood and glucose in urine, glucose in blood). Three papers of Massart et al. [32, 33], the first of which deals with chromatographic separation in a thin layer, are directed to the assessment of the information content of qualitative proofs carried out through the use of chromatography. More extensive is the literature concerning the evaluation of results and methods in quantitative analysis by the use of quantities defined in information theory. These include a report by Kaiser [27] mentioned above, three papers by Doerffel [9, 11, 65], a series of papers by Danzer [28, 29, 40], a study by Grys [66] in which two different analytical methods have been compared, and other publications [2, 3, 21, 22]. Some basic questions have also been discussed in [10]. Various ways of expressing the information content of the results of quantitative analysis have been compared in [12, 14]. In [2, 8, 18, 20, 25, 55], where the information content of the results of a quantitative analysis is expressed by employing the divergence measure, either the normal [2, 18, 55] or the rectangular distribution [20, 25, 55] has been assumed as the a priori distribution, and the normal distribution has always been used as the a posteriori distribution. It has been distinguished whether the analysis results have confirmed [2, 20, 25] the initial rectangular distribution or have not [8], and the assumptions regarding the content of the component to be determined in terms of the expected value of the normal distribution [18, 55]. The information content of the results of quantitative analysis subject to a systematic error has been discussed in [62, 68]. In connection with the possibilities of removing or at least reducing the systematic error of the results of quantitative analysis, papers have been written on the importance of the calibration [69, 70] and its influence upon the information content of the results [16, 68].

The greatest attention in the analytical literature has been drawn to information properties of two-dimensional instrumental methods. These have been treated by Kaiser [27], Danzer [23, 40, 54], Doerffel [9, 11, 65], Gottschalk [42, 50], Zettler [57], Rackow [51], and others [22, 25]. Chromatography, which is a typical example of a two-dimensional analytical method, was perhaps first investigated as an analytical methodology related to the use of information theory by Rackow [51] in 1964 and 1965. Huber and Smith [30], Massart et al. [32, 33], Palm [37], Dupuis and Dijkstra [34], and others [71, 72] have evaluated results and compared various chromatographic methods employing information aspects. Smits et al. [31] have dealt with the optimization of the amount of information obtained by ion-exchange separation and the determination of individual components. The papers of Kaiser [27], Doerffel [11, 65], Danzer [28], and other authors [22, 25, 61] have focused on two-dimensional spectral methods. Grotsch [73] has been concerned with the information content of mass spectra. Automatic data processing is of importance when employing two-dimensional analytical methods. Information aspects in the automatic processing of analytical data have been discussed primarily by Malissa [26] and Gottschalk [42] in the framework of the activities of the "Lindau circle." An exciting survey of the problems of automation in analytical chemistry has been provided in the monograph by Malissa [74]. Information aspects in the consideration of concrete decisions regarding automatic processing of analytical data have also been considered in papers by Seifert [59],

Zettler [57], Kienitz and Kaiser [75], and especially, by Huber and Smith [30]. Grotsch has discussed [73] the possibility of automatically processing the results for mass spectrometry. Van Marlen and Dijkstra [96] and Varmuza and Rotter [97,98] have recently used information contents in evaluations of mass spectra. In identifying organic substances from mass and infrared spectra methods of pattern recognition (cf. Section 6.15) have taken place, based on quantities defined by Shannon's measure (4.2). Information content expressed as a difference of entropies of an a priori distribution and an a posteriori distribution has been used by Dupuis et al. [99] for coding infrared spectra for retrieval purposes.

Problems of the information content of the results of trace analysis have been investigated by Eckschlager [2, 20], later together with Štěpánek [21]. Local analysis carried out, for example, by means of the microprobe correlates closely with the results of trace analysis; it has been evaluated from the point of view of information content by Danzer [28], with the content being considered in the Brillouin sense (i.e., as the logarithm of the number of possible discriminations). Ultramicro elementary analysis has been treated in a very interesting paper by Grieping and Krijgsman [36]. The absolute limits of microchemistry have been discussed by Malissa [46]; Eckschlager [20, 21] has shown that the maximum information content obtainable by chemical or instrumental analysis is correlated with our ability to determine a small amount of a substance analytically.

Analyses are often carried out for quality control needs. Information content and effectivity in analytical quality control have been dealt with rather systematically by Eckschlager [3, 56, 58]. Here the qualitative, and especially the quantitative, side of the information, primarily its relevance, are important for assessing the quality of the product and its content and profitability for evaluation of the effectivity and the economic aspects of the proper control process. Determination of an appropriate probability distribution is especially important for the investigation of the information content of analytical quality control. In [56] two models of a priori probability distributions are considered for the case where the tolerance limits are given. Ehrlich [76] has pointed to the importance of the a priori probability distribution in evaluating the information properties of analytical results. In evaluating the information content in all cases mentioned, the divergence measure has proved very useful. It was employed by Eckschlager and Vajda to evaluate analysis results [18], and later it also appeared to be valuable for other uses [2, 3, 12, 14, 21, 68].

Two papers by Grieping et al. [35, 36] have been devoted to questions of the information content of elementary analysis. The first, published in 1971, shows the practical importance of computing the information contents of determinations of C, H, and N; the importance of the amount of information needed and obtained is also pointed out.

An interesting survey of the historical development of using the information theory in analytical chemistry is presented by Massart et. al. in [100].

References

1. S. Kullback, *Information Theory and Statistics*, John Wiley & Sons, Inc., New York, 1959.
2. K. Eckschlager, *Z. Anal. Chem.* **277**, 1 (1975).
3. K. Eckschlager, *Anal. Chem.* **49**, 1265 (1977).
4. K. Danzer and K. Eckschlager, *Talanta* **25**, 725 (1978).
5. H. Malissa, *Z. Anal. Chem.* **256**, 7 (1971).
6. K. Danzer, *Z. Chem.* **13**, 229 (1973).
7. K. Eckschlager, *Collect. Czech. Chem. Commun.* **36**, 3016 (1971).
8. I. Vajda and K. Eckschlager, *Kybernetika* (in press).
9. K. Doerffel and W. Hildebrand, *Wiss. Z. Tec. Hochsch. Chem. C. Schorlemmer* **11**, 30 (1969).
10. K. Danzer et al., *Analytik*, Akademische Verlagsgesellschaft, Leipzig, 1977.
11. K. Doerffel, *Chem. Tec.* **25**, 94 (1973).
12. K. Eckschlager, *Collect. Czech. Chem. Commun.* **42**, 225 (1977).

13. K. Eckschlager, *Collect. Czech. Chem. Commun.* **43**, 231 (1978).

14. I. L. Larsen, N. A. Hartman, and J. J. Wagner, *Anal. Chem.* **45**, 1511 (1973)

15. K. Eckschlager, *Collect. Czech. Chem. Commun.* **27**, 1521 (1962).

16. K. Eckschlager, *Collect. Czech. Chem. Commun.* **37**, 137 (1972).

17. H. Kaiser, *Z. Anal. Chem.* **209**, 1 (1965); **216**, 80 (1966).

18. K. Eckschlager and I. Vajda, *Collect. Czech. Chem. Commun.* **39**, 3076 (1974).

19. K. Eckschlager, *Collect. Czech. Chem. Commun.* **34**, 1321 (1969).

20. K. Eckschlager, *Collect. Czech. Chem. Commun.* **40**, 3627 (1975).

21. K. Eckschlager and V. Štěpánek, *Mikrochim. Acta* I, 107 (1978).

22. K. Eckschlager, *Z. Chem.* **16**, 111 (1976).

23. K. Danzer, *Z. Chem.* **13**, 69 (1973).

24. K. Doerffel, *Chem. Anal.* **17**, 615 (1972).

25. K. Eckschlager, *Collect. Czech. Chem. Commun.* **41**, 1875 (1976).

26. H. Malissa and J. Rendl, *Z. Anal. Chem.* **272**, 1 (1974); H. Malissa, J. Rendl, and J. L. Marr, *Talanta* **22**, 597 (1975).

27. H. Kaiser, *Anal. Chem.* **42**, No. 2, 24A; (1970).

28. K. Danzer, *Z. Chem.* **15**, 158 (1975).

29. K. Danzer, *Z. Chem.* **14**, 73 (1974).

30. J. F. K. Huber and H. C. Smith, *Z. Anal. Chem.* **245**, 84 (1969).

31. R. Smits, C. Vanroelen, and D. L. Massart, *Z. Anal. Chem.* **273**, 1 (1975).

32. D. L. Massart, *J. Chromatogr.* **79**, 157 (1973).

33. D. L. Massart et al., *Anal. Chem.* **46**, 283, 1988 (1974).

34. F. Dupuis and A. Dijkstra, *Anal. Chem.* **47**, 379 (1975).

35. B. Grieping and G. Dijkstra, *Z. Anal. Chem.* **257**, 269 (1971).

36. B. Grieping and W. Krijgsman, *Z. Anal. Chem.* **265**, 241 (1973).

37. E. Palm, *Z. Anal. Chem.* **256**, 25 (1971).

38. A. Hald, *Statistical Theory with Engineering Applications*, John Wiley & Sons, Inc., New York, 1952.

39. K. Eckschlager, *Collect. Czech. Chem. Commun.* **37**, 1486 (1972).

40. K. Danzer, *Z. Chem.* **15**, 326 (1975).

41. C. Liteanu and J. A. Crisan, *Rev. Roum. Chim.* **12**, 1475 (1967).

42. G. Gottschalk, *Z. Anal. Chem.* **258**, 1 (1972).

43. H. Malissa, *Z. Anal. Chem.* **222**, 100 (1966).

44. H. Malissa and G. Jellinek, *Z. Anal. Chem.* **238**, 81 (1968).

45. H. Malissa, *Z. Anal. Chem.* **247**, 1 (1969).

46. H. Malissa, *Mikrochim. Acta* **1972**, 596.

47. H. Malissa, *Pure Appl. Chem.* **18**, 17 (1969).

48. H. Malissa, *Z. Anal. Chem.* **271**, 97 (1974).

49. H. Kienitz, *Angew. Chem.* **81**, 723 (1969).

50. G. Gottschalk, *Z. Anal. Chem.* **258**, 1 (1972).

51. B. Rackow, *Z. Chem.* **4**, 73, 109, 155 (1964); **5**, 67 (1965).

52. K. Eckschlager, *Collect. Czech. Chem. Commun.* **38**, 1330 (1973).

53. K. Eckschlager, *Collect. Czech. Chem. Commun.*, **39**, 1426 (1974).

54. K. Danzer, *Z. Chem.* **13**, 20 (1973).

55. K. Eckschlager, *Collect. Czech. Chem. Commun.* **42**, 1935 (1977).

56. K. Eckschlager, *Collect. Czech. Chem. Commun.* **43**, 231 (1978).

57. H. Zettler, *Z. Anal. Chem.* **254**, 1 (1961).

58. K. Eckschlager, *Chem. Prum.* **27**, 650 (1977), **28**, 546, 602 (1978).

59. G. Seifert, *Z. Chem.* **11**, 161 (1971).

60. W. Triebal and B. Wilhelmi, *Z. Chem.* **13**, 329 (1973).

61. K. Eckschlager, *CSI Proc.*, Prague, 1977.

62. K. Eckschlager and V. Štěpánek, *Anal. Chem.* (in press).

63. S. N. Deming and S. L. Morgan, *Anal. Chem.* **45**, 278A (1973).

64. S. L. Morgan and S. N. Deming, *Anal. Chem.* **46**, 1170 (1974).

65. S. Arpadjan, K. Doerffel, K. Holland-Letz, H. Much, and M. Pannach, *Z. Anal. Chem.* **220**, 257 (1974).

66. S. Grys, *Z. Anal. Chem.* **273**, 177 (1975).

67. J. Fischer, *Conf. Clin. Chem. Proc.*, Praha 1978.

68. K. Eckschlager, *Collect. Czech. Chem. Commun.* **44**, 111 (1979).

69. H. Kaiser, *Z. Anal. Chem.* **260**, 252 (1972).

70. W. E. van der Linden, *Z. Anal. Chem.* **269**, 26 (1974).

71. A. C. Moffat, A. H. Sted, and K. W. Swaldon, *J. Chromatogr.* **90**, 19 (1974).

72. A. Eskes, F. Dupuis, A. Dijkstra, H. D. De Clerque, and D. L. Massart, *Anal. Chem.* **47**, 168 (1975).

73. S. L. Grotsch, *Anal. Chem.* **42**, 1214 (1970); **45**, 2 (1973).

74. H. Malissa, *Automation in und mit der analytischen Chemie*, Verlag der Wiener Medizinischen Akademie, Vienna, 1972.

75. H. Kienitz and R. Kaiser, *Z. Anal. Chem.* **237**, 241 (1968).

76. G. Ehrlich, *Chem. Anal.* **21**, 303 (1976).

77. W. Spendley, G. R. Hext, and F. R. Hinsworth, *Technometrics* **4**, 441 (1962).

78. G. E. P. Box and K. B. Wilson, *J. R. Stat. Soc.* **B13**, 1 (1951).

79. L. Brillouin, *Scientific Uncertainty and Information*, Academic Press, New York, 1966.

80. L. Brillouin, *Science and Information Theory*, Academic Press, New York, 1962

81. K. Varmuza, *Monatsh. Chem.* **105**, 1 (1974)

82. H. Rotter and K. Varmuza, *Org. Mass. Spectrom* **10**, 874 (1975)

83. P. C. Yurs and T. L. Isenhour, *Chemical Application of Pattern Recognition*, J. Wiley-Interscience, New York, 1975

84. B. G. M. Vandeginste, *Analyt. Letters* **10** (9), 661 (1975)

85. O. Štrouf and J. Fusek, *Coll. Czech. Chem. Commun.* **44**, (1979), in press

86. J. Fusek and O. Štrouf, *Coll. Czech. Chem. Commun.* **44**, (1979), in press

87. S. Wold, *Pattern Recognition* **8**, 127 (1976)

88. H. Abbe and P. C. Yurs, *Anal. Chem.* **47**, 1829 (1975)

89. G. L. Ritter et. al: *Anal. Chem.* **47**, 1951 (1975)

90. K. Varmuza, *Monatsh. Chem.* **107**, 43 (1975); *Z. Anal. Chem.* **266**, 274 (1973), **268**, 352 (1974), **271**, 22 (1974)

91. K. Varmuza, *Z. Anal. Chem.* **286**, 329 (1977)

92. W. E. Brugger, A. Y. Stuper and P. C. Yurs, *Z. Chem. Inf. Comput. Sci.* **16**, 105 (1976)

93. L. Hodes, *Z. Chem. Inf. Comput. Sci.* **16**, 88 (1976)

94. A. Eskes, F. Dupuis, A. Dijkstra, H. D. De Clercg and D. L. Massart, *Anal. Chem.* **47**, 2169 (1975)

95. D. L. Massart and H. D. De Clercg, *Anal. Chem.* **46**, 1988 (1974)

96. G. van Marlen and A. Dijkstra, *Anal. Chem.* **48**, 595 (1976)

97. K. Varmuza and H. Rotter, *Monatsh. Chem.* **107**, 547 (1976)

98. H. Rotter and K. Varmuza, *Adv. Mass. Spectrom.* **7**, 1099 (1978)

99. P.F. Dupuis,. A. Dijkstra and J. H. van der Maas, *Z. Anal. Chem.* **290**, 357, **291**, 27 (1978)

100. D. L. Massart et al: in W. Fresenius, *Reviews on Analytical Chemistry*, Akad. Kiadó, Budapest, 1977.

101. H. C. Andrews, Introduction to Mathematical Techniques in Pattern Recognition, Wiley-Interscience, New York, 1972.

APPENDIX A

A.1 THE χ^2-DISTRIBUTION

This distribution plays an important role in various statistical disciplines. It serves us for instance to derive the sample distribution of sample variances (cf. Section 5.2.2).

Let us use the symbol χ^2 (chi square) to denote the sum of squares of n independent normal standardized random variables

$$\chi^2 = \zeta_1^2 + \zeta_2^2 + \cdots + \zeta_n^2 \qquad (A.1)$$

This sum is distributed with the probability density

$$f_n(x) = \frac{1}{2^{n/2} \cdot \Gamma(n/2)} e^{-x/2} x^{(n/2)-1} \qquad x > 0 \qquad (A.2)$$

where $\Gamma(n/2)$ is the gamma function of the argument $n/2$. We call $f_n(x)$ the probability density of the χ^2-*distribution with n degrees of freedom*.

We obtain for the moments

$$E[\chi^2] = n \qquad V[\chi^2] = 2n \qquad (A.3)$$

The frequency function $f_n(x)$ is positively skew, as seen in Figure A.1. As n increases, the curve approaches symmetry. In the figure χ^2-curves are presented for $n = 2$, 4, and 10.

Percentage points (critical values) $\chi_{n,\alpha}^2$ such that

$$P\{\chi^2 > \chi_{n,\alpha}^2\} = \alpha \qquad (A.4)$$

are tabulated in a number of books on probability and statistics (e.g., in [1]) for various values of n and the most common values of

169

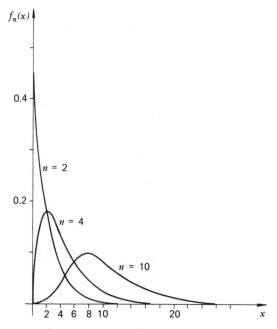

Figure A.1 Frequency function of χ^2-distribution for various values of n.

α. That is, for a fixed area under the curve given by the choice of α, the values in the body of the table give the abscissas $\chi^2_{n,\alpha}$.

A.2 THE STUDENT t-DISTRIBUTION

A second distribution used extensively in statistical tests is the Student t-distribution. This distribution was originally developed by W. S. Gosset under the pseudonym "Student" and later more rigorously by R. A. Fisher.

Consider two independent random variables: ζ is the standardized normal variable and χ^2 has the χ^2-distribution with n degrees of freedom. Then the variable

$$t = \frac{\zeta}{\sqrt{\chi^2/n}} \tag{A.5}$$

can be shown to have the probability density

$$f_n(t) = \frac{1}{\sqrt{n\pi}} \frac{\Gamma\!\left(\dfrac{n+1}{2}\right)^2}{\Gamma(n/2)} \left(1 + \frac{t^2}{n}\right)^{-(n+1)/2} \qquad -\infty < t < \infty$$

$$(A.6)$$

which is known as the probability density of the *Student t-distribution with n degrees of freedom*. As in the χ^2-distribution, this depends only on the single parameter *n*.

Remark.   Denoting the Student variable with *t* of the Latin alphabet has a historical basis. This is true of the random variable in the next section as well.

The *t*-distribution is symmetrical about $t = 0$. Figure A.2 shows this distribution for $n = 4$ compared with the normal curve. It

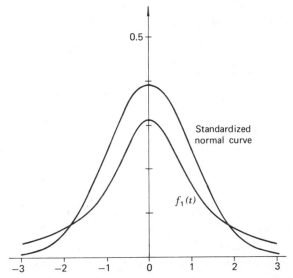

Figure A.2   Frequency function of a Student *t*-distribution compared with the normal distribution.

should be noted that the t-distribution tends to the standard normal distribution as n increases. The variance (for $n > 2$) can be found as

$$V[t] = \frac{n}{n-2}$$

Percentage points (critical values) $t_\alpha(n)$ such that

$$P\{|t| > t_\alpha(n)\} = \alpha \qquad (A.7)$$

are tabulated, for example, in [1] for various values of n and α.

A.3   THE F-DISTRIBUTION

A third distribution which is frequently used in connection with random samples from normal distributions is the F-distribution. It gave rise to an extensive statistical field, including the analysis of variance and experimental designs.

If χ_1^2 and χ_2^2 are independent random variables having χ^2-distributions with n_1 and n_2 degrees of freedom, then

$$F = \frac{\chi_1^2/n_1}{\chi_2^2/n_2} \qquad (A.8)$$

has the probability density

$$f_{n_1,n_2}(F) = \frac{\Gamma\left(\dfrac{n_1+n_2}{2}\right)}{\Gamma(n_1/2)\Gamma(n_2/2)} \left(\frac{n_1}{n_2}\right)^{n_1/2} F^{(n_1/2)-1}\left(1 + \frac{n_1}{n_2}F\right)^{-(n_1+n_2)/2}$$

$$(A.9)$$

for $F > 0$, which is called the probability density of the F-distribution with n_1 and n_2 degrees of freedom. Thus, this is a two-parameter family of distributions.

Percentage points (critical values) $F_\alpha(n_1, n_2)$ such that

$$P\{F > F_\alpha(n_1, n_2)\} = \alpha \qquad (A.10)$$

for various n_1 and n_2 and for the most common values of α are tabulated, for example, in [2].

The F-distribution with a pair of degrees of freedom $(1, n)$ is identical with the distribution of the random variable t^2. Therefore, the critical values $F_\alpha(1, n)$ are the squares of $t_\alpha(n)$.

References

1. E. S. Pearson and H. O. Hartley, *Biometrika Tables for Statisticians*, Vol. 1, Cambridge University Press, New York, 1958.
2. J. A. Greenwood and H. O. Hartley, *Guide to Tables in Mathematical Statistics*, Princeton University Press, Princeton, N.J., 1961.

APPENDIX B

Several results, such as values of information contents and the errors in the calibration of analytical methods, shown in sections of Chapter 6 were derived by the use of the mathematical apparatus presented in Chapters 3, 4, and 5. Since the way of deriving various information measures is both instructive for the reader in his attempts to carry out calculations in other cases and shows more deeply the adoption of probabilistic technique and information theory concepts, a few derivations will be illustrated in this part B of the Appendix. Moreover statistical analysis in sections on calibration will be clarified here.

B.1 INFORMATION CONTENT OF A QUALITATIVE PROOF

Assuming the number of components prior to the analysis and after the analysis to be k_0 and k, respectively (see Section 6.2), we get for the information content using (4.7) for discrete uniform distributions according to the remark on page 67

$$I(p,p_0) = \sum_{i=1}^{k} \frac{1}{k} \ln \frac{\frac{1}{k}}{\frac{1}{k_0}} = k \cdot \frac{1}{k} \ln \frac{k_0}{k} = \ln \frac{k_0}{k}$$

which is the result shown in (6.1).

B.2 INFORMATION CONTENT OF A QUANTITATIVE DETERMINATION

The a priori probability distribution of our knowledge of the true value of the component to be determined is uniform on an interval $\langle x_1, x_2 \rangle$. The probability distribution of the results of the analysis is supposed to be normal and fulfilling the condition $\mu \in \langle x_1 + 3\sigma, x_2 - 3\sigma \rangle$ where μ and σ are the mean value and the standard deviation, respectively. Then, according to (4.12), the information content is calculated as

$$I(p, p_0)$$

$$= \int_{x_1}^{x_2} \frac{1}{\sqrt{2\pi}\,\sigma} \exp\left[-\frac{1}{2}\left(\frac{x-\mu}{\sigma}\right)^2\right] \cdot \ln \frac{(x_2 - x_1) \exp\left[-\frac{1}{2}\left(\frac{x-\mu}{\sigma}\right)^2\right]}{\sqrt{2\pi}\,\sigma}\, dx$$

$$= \ln \frac{x_2 - x_1}{\sqrt{2\pi}\,\sigma} \int_{x_1}^{x_2} \frac{1}{\sqrt{2\pi}\,\sigma} \exp\left[-\frac{1}{2}\left(\frac{x-\mu}{\sigma}\right)^2\right] dx$$

$$- \frac{1}{2\sigma^2} \int_{x_1}^{x_2} \frac{(x-\mu)^2}{\sqrt{2\pi}\,\sigma} \exp\left[-\frac{1}{2}\left(\frac{x-\mu}{\sigma}\right)^2\right] dx$$

However, under the assumption above (more than 99.7% of the results lie within the interval $\langle x_1, x_2 \rangle$), the first integral can be approximated by the unity and the second one approximates, according to the definition in (3.19), the variance σ^2. Thus

$$I(p, p_0) = \ln \frac{x_2 - x_1}{\sqrt{2\pi}\,\sigma} - \frac{1}{2\sigma^2} \cdot \sigma^2 = \ln \frac{x_2 - x_1}{\sigma\sqrt{2\pi e}}$$

which refers to (6.3).

B.3   INFORMATION CONTENT IN HIGHER-PRECISION ANALYSIS

If a quantitative analysis is repeated by the use of another, more precise, method (Section 6.6), two normal distributions ($\sigma_0 > \sigma$) enter the formula for the information content in (4.12). Thus, we obtain

$$I(p,p_0)=\int_{-\infty}^{\infty}\frac{1}{\sqrt{2\pi}\,\sigma}\left[\ln\frac{\sigma_0}{\sigma}-\frac{1}{2}\left\{\left(\frac{x-\mu}{\sigma}\right)^2-\left(\frac{x-\mu_0}{\sigma_0}\right)^2\right\}\right]$$

$$\exp\left[-\frac{1}{2}\left(\frac{x-\mu}{\sigma}\right)^2\right]dx$$

$$=\ln\frac{\sigma_0}{\sigma}-\frac{1}{2}+\frac{1}{2\sigma_0^2}\int_{-\infty}^{\infty}\frac{(x-\mu_0)^2}{\sqrt{2\pi}\,\sigma}\exp\left[-\frac{1}{2}\left(\frac{x-\mu}{\sigma}\right)^2\right]dx$$

because the first two integrals equal the unity and the variance σ^2, respectively, similarly as in the Section B.2. The last integral can be manipulated into the form

$$\int_{-\infty}^{\infty}\left[(x-\mu)+(\mu-\mu_0)\right]^2\frac{1}{\sqrt{2\pi}\,\sigma}\exp\left[-\frac{1}{2}\left(\frac{x-\mu}{\sigma}\right)^2\right]dx$$

$$=\int_{-\infty}^{\infty}\frac{(x-\mu)^2}{\sqrt{2\pi}\,\sigma}\exp\left[-\frac{1}{2}\left(\frac{x-\mu}{\sigma}\right)^2\right]dx+(\mu-\mu_0)^2\cdot 1$$

$$=\sigma^2+(\mu-\mu_0)^2$$

as the middle term vanishes (the central moment of the first order is always equal to zero). And so

$$I(p,p_0)=\ln\frac{\sigma_0}{\sigma}-\frac{1}{2}+\frac{1}{2\sigma_0^2}\left[\sigma^2+(\mu-\mu_0)^2\right]$$

$$=\ln\frac{\sigma_0}{\sigma}+\frac{(\mu-\mu_0)^2+\sigma^2-\sigma_0^2}{2\sigma_0^2}$$

which is the result shown in (6.10).

B.4 INFLUENCE ON INFORMATION CONTENT OF ACCURACY OF ANALYTICAL METHODS

If the requirement of the accuracy (unbiasness) of the results of quantitative determinations is not fulfilled, that is if the results are subject to a systematic error $\delta=|X-E[\xi]|$ (see Section 6.4), the information content is lower than is given in (6.3).

Let us assume again a uniform a priori distribution in the interval $\langle x_1, x_2 \rangle$ and let $p_1(x)$ be a normal distribution around the true value X and let $p_2(x)$ be a normal distribution with the mean value $\mu = E[\xi]$ (both with the same standard deviation σ). If the employed analytical method is biased, we can measure the information content as a difference

$$I(p_2, p_1 / p_1, p_0) = I(p_1, p_0) - I(p_2, p_1)$$

The first integral can be evaluated according to (6.3) and the value of the second one can be obtained from (6.10) by substituting $\sigma_0 = \sigma$ and $\mu - \mu_0 = \delta$ and it turns out $\frac{1}{2} \left(\frac{\delta}{\sigma} \right)^2$, so that

$$I(p_2, p_1 / p_1, p_0) = \ln \frac{x_2 - x_1}{\sigma \sqrt{2\pi e}} - \frac{1}{2} \left(\frac{\delta}{\sigma} \right)^2$$

which is the result shown in (6.6). Thus, if an analytical method is subject to a systematic error, the information content is always lower than was found in (6.3).

B.5   INFLUENCE ON INFORMATION CONTENT OF CALIBRATION OF ANALYTICAL METHODS

According to (6.7a), the result of a quantitative analysis that is related to the readout by a proportionality factor, when calibration is carried out by a standard addition, is obtained as

$$x_1 = \frac{y_1}{y_s} x_s = x_s \frac{y_1}{y_2 - y_1}$$

where $y_2 > y_1$.

Thus, x_1 is a function of values of random variables y_1 and y_2 and its variance can be approximated by

$$\sigma_{x_1}^2 \approx \left[\frac{\partial x_1(\mu_{y_1}, \mu_{y_2})}{\partial y_1} \right]^2 \sigma_{y_1}^2 + \left[\frac{\partial x_1(\mu_{y_1}, \mu_{y_2})}{\partial y_2} \right]^2 \sigma_{y_2}^2 \qquad (B.1)$$

This approximation follows from replacing an arbitrary function with the linear terms of its Taylor series expansion and calculating its variance by the use of theorems on variance. Passing to the sample variance we get, after having calculated the partial derivatives (with one observation y_1 and y_2 each), and assuming that the standard deviation is independent of the measured value

$$s_{x_1}^2 \approx s_y^2 \cdot x_s^2 \frac{y_1^2 + y_2^2}{(y_2 - y_1)^4}$$

and for the sample standard deviation

$$s_{x_1} \approx \frac{s_y \cdot x_s}{(y_2 - y_1)^2} \sqrt{y_1^2 + y_2^2}$$

cf. (6.76).

In the case of constructing a calibration straight line, as discussed in Section 6.5, by the method of linear regression outlined in Section 5.5, we need to find confidence limits for the result of a determination provided that the empirical calibration straight line was constructed of m points. Since the intercept a in the regression function can be replaced according to (5.21), we can write the regression function as

$$y = \bar{y} + b(x - \bar{x})$$

Calculating a value x_{m+1} corresponding to intensity $y = y_{m+1}$ in analyzing this sample, we get

$$x_{m+1} = \bar{x} + \frac{y_{m+1} - \bar{y}}{b}$$

Hence, it is a function of three random variables y_{m+1}, $\bar{y}$, and b and its variance can be approximated similarly as in (B.1). After computing the partial derivatives we obtain for the sample variance

$$s_{x_{m+1}}^2 = \frac{1}{b^2}\left(s_{y_{m+1}}^2 + s_{\bar{y}}^2\right) + \frac{\left(y_{m+1}-\bar{y}\right)^2}{b^4}s_b^2$$

where $s_{y_{m+1}}^2$ is the variance $s_{y \cdot x}^2$ about the regression line given in (5.22), $s_{\bar{y}}^2 = (1/m)s_{y \cdot x}^2$ and s_b^2 is the estimate of $V[b]$ (see page 101). In this way the error in the determination of the unknown concentration x_{m+1} will be given by the standard deviation

$$s_{x_{m+1}} = \frac{s_{y \cdot x}}{b}\sqrt{1 + \frac{1}{m} + \frac{\left(y_{m+1}-\bar{y}\right)^2}{b^2 \sum_i \left(x_i - \bar{x}\right)^2}}$$

And the $100(1-\alpha)$ percent confidence limits are found as

$$x_{m+1} \pm t_\alpha(m-2)s_{x_{m+1}}$$

We observe that the standard deviation of the determination of the unknown concentration assumes its minimum when $y_{m+1}=\bar{y}$. In this particular case the error in the slope of the calibration line has no influence upon the result of the analysis and its standard deviation

$$s_{x_{m+1}} = \frac{s_{y \cdot x}}{b}\sqrt{\frac{m+1}{m}}$$

If the unknown sample is analyzed by n_p parallel determinations, then the error in finding the unknown concentration expressed in terms of the standard deviation appears as

$$s_{\bar{x}(n_p)} = \frac{s_{y \cdot x}}{b}\sqrt{\frac{1}{n_p} + \frac{1}{m} + \frac{\left(\bar{y}(n_p)-\bar{y}\right)^2}{b^2 \sum_i \left(x_i - \bar{x}\right)^2}}$$

where $\bar{y}(n_p) = \frac{1}{n_p}\sum_{i=1}^{n_p} y_{m+i}$, which is the result from (6.7d) (except for symbols used in Section 6.5).

B.6   INFORMATION CONTENT OF
TRACE-ANALYSIS RESULTS

We will denote by x_0 the determination limit, that is the least content determinable by the given method. The distribution of results of a trace analysis that only indicate that the content of the component to be determined is less than the determination limit is supposed to be rectangular

$$p(x) = \begin{cases} \dfrac{1}{x_0} & \text{for } x \in \langle 0, x_0 \rangle \\ 0 & \text{otherwise} \end{cases}$$

If the highest expected content is $x_1 > x_0$ and no preference is given beforehand to any value of the interval $\langle 0, x_1 \rangle$, we obtain for the divergence information measure

$$I(p,p_0)_1 = \int_0^{x_0} \frac{1}{x_0} \ln \frac{\frac{1}{x_0}}{\frac{1}{x_1}} \, dx = \ln \frac{x_1}{x_0}$$

which is the result shown in (6.14).

If the content of the component to be determined is greater than x_0, its distribution is assumed to be log-normal of the form

$$p(x) = \begin{cases} \dfrac{1}{(x-x_0)\sigma\sqrt{2\pi}} \exp\left\{ -\frac{1}{2} \left[\frac{\ln(x-x_0) - \mu}{\sigma} \right]^2 \right\} & x \in (x_0, \infty) \\ 0 & \text{otherwise} \end{cases}$$

where μ and σ are parameters of the normal distribution of the logarithms $\ln(x - x_0)$, in which we can set $\mu = \ln kx_0$, $k > 0$. Substituting in the formula for the information divergence measure, we obtain

$$I(p,p_0)_2 = \int_{x_0}^{x_1} p(x) \left\{ \ln \frac{x_1}{\sqrt{2\pi}\,\sigma} - \ln(x-x_0) - \frac{1}{2} \left[\frac{\ln(x-x_0) - \mu}{\sigma} \right]^2 \right\} dx$$

The first integral equals $\ln \dfrac{x_1}{\sqrt{2\pi}\,\sigma}$ after an approximation

$$\int_{x_0}^{x_1} p(x)\,dx \approx 1$$

Substituting

$$\ln(x - x_0) = t, \quad \frac{dx}{x - x_0} = dt$$

in the second integral, we get it in the form

$$J_2 = \int_{-\infty}^{\ln(x_1 - x_0)} t \cdot \frac{1}{\sqrt{2\pi}\,\sigma} \exp\left[-\frac{1}{2}\left(\frac{t-\mu}{\sigma}\right)^2 \right] dt \approx \mu$$

by the definition of the mean value in (3.17), for $\ln(x_1 - x_0)$ can be assumed to be sufficiently large. Similarly, by the same substitution, the third integral appears as

$$J_3 = \frac{1}{2\sigma^2} \int_{-\infty}^{\ln(x_1 - x_0)} (t - \mu)^2 \frac{1}{\sqrt{2\pi}\,\sigma} \exp\left[-\frac{1}{2}\left(\frac{t-\mu}{\sigma}\right)^2 \right] dt \approx \frac{1}{2\sigma^2} \cdot \sigma^2$$

$$= \frac{1}{2}$$

by the definition of the variance in (3.19). Hence the information content

$$I(p,p_0)_2 = \ln \frac{x_1}{\sqrt{2\pi}\,\sigma} - \mu - \frac{1}{2}$$

$$= \ln \frac{x_1}{\sigma\sqrt{2\pi e}} - \mu$$

and after setting $\mu = \ln kx_0$

$$I(p,p_0)_2 = \ln\left(\frac{x_1}{x_0} \cdot \frac{1}{k\sigma\sqrt{2\pi e}} \right)$$

which is the value shown in (6.15) (except for $\sqrt{n_p}$ not applicable in this derivation).

Remark. The validity of the used approximations is in practice guaranteed. Although x_1 is taken as a maximum assumed value of the component to be determined, the values of the log-normal random variable used in the model extend to infinity. However, the frequency function of this distribution is in most cases strongly assymetrical with the mode close to x_0. If we require, for example, that the logarithms of the concentrations exceed the value $\ln(x_1 - x_0)$ in at most 5 percent cases, it follows from the inequality

$$\ln(x_1 - x_0) - \ln kx_0 > 1.65\sigma$$

that

$$k < \left(\frac{x_1}{x_0} - 1\right)\exp(-1.65\sigma)$$

or that the mean value of the log-normal distribution fulfill the inequality

$$E[\xi] < x_0 + (x_1 - x_0)\exp\left(\frac{1}{2}\sigma^2 - 1.65\sigma\right)$$

INDEX